India's Energy Revolution

India is the third-largest emitter of greenhouse gases, which makes it an important player whose climate mitigation actions and inactions are closely scrutinised. This book studies developments in India's energy system from a governance perspective. It presents a unique compilation and synthesis of research findings that capture achievements, shortcomings, and persistent and transient challenges of India's transition towards a net-zero economy by 2070.

The book grounds its analysis in domestically formulated goals and reflects on dynamics at the structural level of India's multi-scalar innovation system, by highlighting the influencing factors of energy system status and change. It presents the perspectives and positions of different actor groups, studies the market and business, and discusses cases influenced by existing or changing institutions across the whole spectrum of energy resources from fossil to non-fossil fuels and respective technologies.

The volume will be useful for students and researchers in energy governance, energy policy and economics, socio-technical transition studies, energy systems engineering, sustainable development, and environmental studies. It will also be of interest to policymakers and investors.

Annika Bose Styczynski is associate professor and associate dean (research and international collaboration) at the Jindal School of Government and Public Policy, Sonipat, India.

India's Energy Revolution

Insights into the Becoming of a Global Power

Edited by Annika Bose Styczynski

Routledge
Taylor & Francis Group
LONDON AND NEW YORK

First published 2024

by Routledge
4 Park Square, Milton Park, Abingdon, Oxon OX14 4RN

and by Routledge
605 Third Avenue, New York, NY 10158

Routledge is an imprint of the Taylor & Francis Group, an informa business

British Library Cataloguing-in-Publication Data
A catalogue record for this book is available from the British Library

ISBN: 978-1-032-25152-3 (hbk)
ISBN: 978-1-032-25154-7 (pbk)
ISBN: 978-1-003-28181-8 (ebk)

DOI: 10.4324/9781003281818

Typeset in Sabon
by Apex CoVantage, LLC

Contents

Figures

Tables

Contributors

Aayushi Agarwal, a Triple Gold Medallist from IIM-Bangalore/Delhi College of Engineering, currently serves as a senior business analyst at Pricewaterhouse Coopers. With diverse energy sector experience, her previous roles included strategic planning, business development, process re-engineering, and internal audit at Pipeline Infrastructure Ltd, market analysis at Reliance Industries, and operational management at the National Thermal Power corporation (NTPC) and Nuclear Power Corporation of India Ltd.

Kaveri Ashok, a senior associate at Centre for Study of Science, Technology and Policy (CSTEP), holds a BSc in physics and a dual degree in nuclear science and technology. With experience at the French Atomic Energy Commission and research roles at the Observer Research Foundation and the National Institute of Advanced Studies, Bangalore, she co-built Sustainable Alternative Futures for India (SAFARI), a system dynamics model for alternative energy scenarios in India. Currently, she focuses on developing net-zero pathways for India.

Abhijit Banerjee, PhD, is a professor of environmental studies at the Jindal School of Liberal Arts and Humanities with over 10 years of work experience in the United States. He has played a key role in sustainability initiatives, collaborated with AASHE, and contributed to environmental policy in India through GIZ and the United Nations. His expertise includes environmental and energy policy, resource efficiency, and circular economy.

Srestha Banerjee, PhD, is a public policy expert and the director of the India Just Transition Centre at iFOREST. With over 12 years of international work experience, her focus is on climate change and Just Transition in the mining sector and participatory governance. She advises national- and state-level institutions and serves on the advisory board of the Just Transition Initiative.

Tirtha Biswas, a policy analyst, focuses on sustainable pathways for India's industry growth with low-carbon solutions. With a background in mineral engineering, he now leads policy advocacy and strategic partnerships for SunEdison, aiming to establish India as a global hub for electrolyser manufacturing and make green hydrogen globally affordable.

Chandra Bhushan, the president and CEO of iFOREST, is a prominent Indian environment and climate change expert. Formerly with the Centre for Science and Environment, he has authored 20 books and led impactful studies on food safety and environmental pollution. Advising global institutions, he co-chaired the Global Reporting Initiative's Technical Advisory Committee and serves as an expert member for the Brazil, South Africa, India, and China group on climate change.

Sanjay Chaudhury, an associate professor of strategy at Jindal Global Business School, focuses on entrepreneurial orientation, absorptive capacity, and small family businesses. He has 20 years of industry experience in automotive, welding, and consumer durables, and holds a fellowship in strategic management from IIM Kashipur, and his research is published in esteemed international journals.

Rishu Garg is a seasoned professional with more than a decade of experience in the energy and power sector. As group lead at CSTEP, she focuses on developing progressive policies for sustainable clean energy transformation. With an MBA in power management, her expertise includes strategy, planning, clean energy transition, and distribution reforms in collaboration with electricity utilities.

G.V. Hanumanth Raju is a senior analyst at CSTEP. He works on demand–supply analysis for utilities and implementation of distribution reforms for Karnataka utilities. He has completed his MTech in power system engineering and a bachelor's degree in electrical and electronics engineering from Visvesvaraya Technological University, Belagavi, Karnataka. His area of interest is power system analysis.

A.T. Ajmal Khan, PhD, a researcher at Harvard's South Asia Institute, specialises in development, environment, and democratic politics. His PhD focused on India's Nuclear Expansion policy. With experience at the School of Oriental and African Studies, University of London, and Ashoka University, his work appears in journals like *South Asia Research* and *Economic and Political Weekly*.

Sarah Khan, a senior analyst at CSTEP, is pursuing her PhD in air pollution modelling. With an MTech in environmental engineering, she leads net-zero pathway development for India using the SAFARI model. Formerly head of department at K.R. Mangalam University, her expertise includes environmental modelling, system dynamics, and air quality analysis, with notable projects in drone-based air quality monitoring and green concrete.

Kishore Kumar Gangwani, PhD, is an assistant professor of marketing at MDI Gurgaon with a PhD from IIM Bangalore. His research focuses on consumer decision-making, offloading consumer experiences, and digital memories, with publications in esteemed journals like the *Journal of Cleaner Production* and *International Journal of Retail and Distribution Management*.

E.V. Mallik is a senior analyst at CSTEP with extensive experience in power sector research. Holding a bachelor's degree in electrical engineering, he excels in energy policy, demand forecast, and grid planning studies. His strong analytical and project management skills make him a successful cross-functional collaborator, delivering projects with excellence.

Akhil Mehrotra, PhD, is the CEO of Pipeline Infrastructure Limited, a Brookfield company, with more than three decades of experience in energy and infrastructure. Formerly at Shell for 15 years, he is a recognised thought leader and shaped Indian oil and gas market regulations. He authored the "Vision 2030" report. He is a trained mechanical engineer, has a master's in finance and a PhD in gas markets.

Ramya Natarajan, a research scientist at CSTEP Bangalore, is leading the Climate Change Mitigation Group. With a master's degree in environmental engineering from Johns Hopkins University, she has 12 years of interdisciplinary research experience in energy modelling, microbiology, and biofuels. She developed the SAFARI model and contributed to projects evaluating biofuels for India's alternative fuels pathway.

Parvej, the COO at eOxigen Automotive, holds a master's degree in public policy from Jindal School of Government and Public Policy and a BA in technology from Indian Institute of Technology (IIT) Delhi. He has diverse experience in clean energy solutions, contributing to organisations like the International Solar Alliance, Organisation for Economic Cooperation and Development, Council on Energy, Environment and Water (CEEW), and MASS Foundation, focusing on mobility and social sustainability.

Deepak Sangroya, PhD, is an associate professor of marketing at Jindal Global Business School, O.P. Jindal Global University. His research focuses on green marketing, B2B marketing, sustainability, and corporate environmental initiatives. He has published in journals such as the *Journal of Strategic Marketing*, *Journal of Cleaner Production*, and *Organization and Environment*.

Debmalya Sen, an energy professional with more than a decade of work experience, including at KPMG and Customised Energy Solutions, currently serves as India Lead – Energy at the World Economic Forum. His previous roles included control room operations, asset management, project management, and innovation. With a diverse background in coal, gas, renewables, and energy storage, he holds degrees in instrumentation, thermal power plant engineering, and an MBA with a finance specialisation from SP Jain School of Global Management.

Rahul Walawalkar, PhD, leads Customised Energy Solutions' emerging technologies and markets practice, specialising in energy storage, renewables, demand response, and smart grid technologies. With extensive board and leadership roles, he is a key figure in global energy storage initiatives, contributing to policy development in the United States and India.

Deepak Yadav, PhD, a policy researcher at CEEW, specialises in low/no carbon fuels for industrial decarbonisation. With degrees from IIT Bombay, he focuses on developing road maps for hard-to-decarbonise industrial sectors, emphasising alternative fuels and innovative technologies to address the challenges of carbon reduction.

Acronyms

AAC	Autoclaved Aerated Concrete
AC	Alternating Current
A/C	Air Conditioner
ACS	Average Cost of Supply
AD	Accelerated Depreciation
AE	Alkaline Electrolyser
AMRUT	Atal Mission for Rejuvenation and Urban Transformation
APDP	Accelerated Power Development Programme
APDRP	Accelerated Power Development and Reforms Programme
ARAI	Automobiles Research Association of India
ARR	Average Revenue Realisation
AT&C	Aggregate Technical and Commercial
BARC	Bhabha Atomic Research Centre
BAU	Business as Usual
BCB	Burnt Clay Bricks
BEE	Bureau of Energy Efficiency
BEV	Battery Electric Vehicle
BESS	Battery Energy Storage System
BG	British Gas
BMI	Business Model Innovation
BMS	Behind-the-Meter Storage
BoS	Balance of Storage
BWR	Boiling Water Reactor
CAPEX	Capital Expenditures
CCS	Carbon Capture and Storage
CEA	Central Electricity Authority
CEEW	Council on Energy, Environment and Water
CERC	Central Electricity Regulatory Commission
CES	Customised Energy Solutions
CFA	Central Financial Assistance
CG	Coal Gasification
CIL	Coal India Limited
CNG	Compressed Natural Gas

CLD	Causal Loop Diagram
CMPDI	Central Mine Planning and Design Institute
CO_2	Carbon Dioxide
CO_2-eq	Carbon Dioxide Equivalent
COP21	21st Conference of the Parties in Paris in 2015.
COP26	26th Conference of the Parties in Glasgow in 2021
CORE	Central Organization for Railway Electrification
Crore	Short form of 10 million in the Indian numbering system.
CSCAF	Climate Smart Cities Assessment Framework
CSR	Corporate Social Responsibility
DAE	Department of Atomic Energy
DC	Direct Current
DGH	Director General of Hydrocarbons
DISCOM	Power Distribution Company
DMF	District Mineral Foundation
DSM	Demand-Side Management
DT	Distribution Transformer
EBP	Ethanol-Blended Petrol
ECBC	Energy Conservation Building Code
EESL	Energy Efficiency Services Limited
EIA	Energy Information Administration
EMS	Energy Management System
EPA	Environmental Protection Agency
ESS	Energy Storage System
EU	European Union
EV	Electric Vehicle
EVI	Electric Vehicle Initiative
FaL-G	Fly Ash-Lime-Gypsum
FAME	Faster Adoption and Manufacturing of Electric Vehicles
FBR	Fast Breeder Reactor
FCEV	Fuel Cell Electric Vehicle
FCI	Food Corporation of India
FDI	Foreign Direct Investment
FERC	Federal Energy Regulatory Commission
FIT	Feed-In Tariff
FRA	Forest Rights Act
FSI	Floor Space Index
GBI	Generation-Based Incentive
GDP	Growth Domestic Product
GENCO	Generation Company
GHG	Greenhouse Gas
GoI	Government of India
GST	Goods and Services Tax
GW	Gigawatt
GWEC	Global Wind Energy Council

HAREDA	Haryana Renewable Energy Development Agency
HELP	Hydrocarbon Exploration and Licensing Policy
HIG	Higher-Income Group
HRS	Hydrogen Refuelling Station
HP	Horsepower (hp) is a unit used to express the rate at which mechanical energy is expanded. A power level of 1 hp is approximately equivalent to 0.746 kilowatts (KW).
HPCL	Hindustan Petroleum Corporation Limited
ICCT	International Council on Clean Transportation
ICE	Internal Combustion Engine
ICT	Information and Communication Technology
IEA	International Energy Agency
IESA	India Energy Storage Alliance
IEX	Indian Energy Exchange Limited
iFOREST	International Forum on Environment, Sustainability and Technology
IIT	Indian Institute of Technology
ILUC	Indirect Land-Use Change
IMF	International Monetary Fund
InvIT	Investment Infrastructure Trust
IOC	Indian Oil Corporation
IPCC	Intergovernmental Panel on Climate Change
IPP	Independent Power Producer
IREDA	Indian Renewable Energy Development Agency
IRENA	International Renewable Energy Agency
IRR	Internal Rate of Return
ITUC	International Trade Union Confederation
Lakh	Short form of 100,000 in the Indian numbering system.
LCOS	Levelised Cost of Storage
LCV	Light Commercial Vehicles
LNG	Liquefied Natural Gas
LOHC	Liquid Organic Hydrogen Carriers
MEA	Ministry of External Affairs
MGNREGA	Mahatma Gandhi National Rural Employment Guarantee Act
MIG	Middle-Income Group
MMBTU	One Million British Thermal Units
MMDRA	Mines and Minerals Development and Regulation Act
MMTPA	Metric Million Tonne per Annum
MNRE	Ministry of New and Renewable Energy
MoHUA	Ministry of Housing and Urban Affairs
MoEFCC	Ministry of Environment, Forests, and Climate Change
MoF	Ministry of Finance
MoP	Ministry of Power
MoPNG	Ministry of Petroleum and Natural Gas

MOSPI	Ministry of Statistics and Programme Implementation
MSW	Municipal Solid Waste
MTPA	Million Tonne Per Annum
MW	Megawatt
NBC	National Building Code
NBCC	National Biofuel Coordination Committee
NBM	National Biodiesel Mission
NCAP	National Cooling Action Plan
NCDC	National Coal Development Corporation of India Ltd.
NDC	Nationally Determined Contribution
NELP	National Exploration and Licensing Policy
NEMMP	National Electric Mobility Mission Plan
NISE	National Institute of Solar Energy
NIUA	National Institute of Urban Affairs
NLC	Neyveli Lignite Corporation Limited
NPB	National Policy on Biofuels
NPCIL	Nuclear Power Corporation of India Limited
NTPC	National Thermal Power Corporation
OALP	Open Acreage Licensing Policy
OCED	Organisation for Economic Cooperation and Development
OEM	Original Equipment Manufacturer
OMC	Oil Marketing Companies
OPEX	Operational Expenditures
PCS	Power Control System
PEM	Proton Exchange Membrane
PF	Power Factor
PFC	Power Finance Corporation
PHWR	Pressurised Heavy Water Reactor
PLI	Production Linked Incentive
PMP	Phased Manufacturing Programme
PNGRB	Petroleum and Natural Gas Regulatory Board
POSOCO	Power System Operation Corporation Limited is a wholly owned Public Sector Undertaking of Government of India under Ministry of Power
PPCA	Powering Past Coal Alliance
PPA	Power Purchase Agreement
PPP	Public–Private Partnership
PRI	Panchayati Raj Institution
PSC	Production Sharing Contract
PV	Photovoltaic
PWR	Pressurised Water Reactor
PXIL	Power Exchange India Limited
R-LNG	Regasified Liquefied Natural Gas
RA	Reverse Auctions
R&D	Research and Development

RE	Renewable Energy
REC	Renewable Energy Certificates
RED	Renewable Energy Directive
REDA	Renewable Energy Development Authority
RFS	Renewable Fuel Standard
RMI	Rocky Mountains Institute
RPO	Renewable Purchase Obligation
RTPV	Rooftop Photovoltaic
SAF	Sustainable Aviation Fuel
SAFARI	Sustainable Alternative Futures for India
SDG	Sustainable Development Goal
SDR	Special Drawing Rights
SDS	Sustainable Development Scenario
SEB	State Electricity Board
SECI	Solar Energy Corporation of India
SERC	State Electricity Regulatory Commission
SIAM	Society of Indian Automobile Manufacturers
SPM	Suspended Particulate Matter
STEPS	Stated Policies Scenario
T&D	Transmission & Distribution
TAPS	Tarapur Atomic Power Station
TEDA	Tamil Nadu Energy Development Agency
TERI	The Energy Resources Institute of India
TIFAC	Technology Information, Forecasting, and Assessment Council
TOD	Transit-Oriented Development
TPP	Thermal Power Plant
TSF	Transformative
TPD	Tonne per Day
TRANSCO	Transmission Company
UCO	Used Cooking Oil
UHI	Urban Heat Island
ULB	Urban Local Bodies
UNFCCC	United Nations Framework Convention on Climate Change
VGF	Viability Gap Funding

1 India's Energy (R)evolution

Annika Bose Styczynski

Fossil fuels started to rapidly expand since the 1950s, taking up roughly 80 per cent of global energy consumption in 2019 (Ritchie, 2019). Today, the speed at which the global energy system must be restructured and unlocked from its fossil fuel dependence is nothing short of a revolutionary endeavour. In fact, to remain within the 1.5°C temperature increase above pre-industrial levels – determined to maintain our ecological integrity – would mean bringing down global anthropogenic Greenhouse Gas (GHG) emissions to zero by 2032 (Figueres et al., 2017). As a developing country covered by the principle of Common but Differentiated Responsibilities and Respective Capabilities, India has legitimately become the third-largest emitter of GHGs. With that, India has also become an important player whose climate mitigation actions and inactions are closely scrutinised. India's economic growth, industrialisation efforts, population growth, and urban drift put the country at the global top of the International Energy Agency's (IEA's) energy demand growth projections until 2040 (IEA, 2021). India's energy use has doubled since the year 2000 and to meet the expected electricity demand of the next 20 years, it will need to install an additional capacity of roughly 850 gigawatts (GW) – the magnitude of the European Union's (EU's) power system.

With "India's Energy (R)evolution," we present a unique compilation and synthesis of research findings that capture achievements, shortcomings, and persistent and transient challenges of India's energy system change towards a net-zero economy by 2070. We ground our analysis in domestically formulated goals and reflect on dynamics at the structural level of India's multi-scalar innovation system, by highlighting the influencing factors of energy system status and change. This includes perspectives and positions of different actor groups and how they interact, the market, business, and use cases influenced by existing or changing institutions across the whole spectrum of energy resources from fossil to non-fossil fuels and respective technologies.

External and peer pressures as much as the ambition to show global leadership have moved the Government of India into international commitments and placed India in the group of solution-oriented countries. The first set of Nationally Determined Contributions (NDCs) to the United Nations Framework Convention on Climate Change (UNFCCC) and the revised set of

DOI: 10.4324/9781003281818-1

Table 1.1 India's NDCs of 2015 and 2021.

2015 *First set of NDCs*	*2021* *Second set of NDCs*
1. Reduction of GHG emissions intensity per unit of GDP by 33–35% by 2030 (base year 2005)	1. About 45% GHG emissions intensity reduction per unit of GDP by 2030 (base year 2005) 2. Reduction of emissions by one billion tonnes by 2030
2. Increase of non-fossil fuel power capacity to 40% by 2030 (conditional on technology transfer and finance)	3. Increase of non-fossil fuel power capacity to 50% by 2030 4. Increase of installed power capacity to 500 GW from renewable sources of energy
3. Additional carbon sink of cumulative 2.5–3.0 Gt CO_2-eq by 2030 through additional forest and tree cover	5. Reach carbon net-zero by 2070
	"One-World Movement" for LIFE – a Lifestyle for Environment

Note: CO_2-eq, carbon dioxide equivalent; GDP, gross domestic product.

Source: UNFCCC (2015) and HT (2021)

NDCs announced by Prime Minister Modi at the 26th Conference of the Parties (COP26) in Glasgow in November 2021 are strong testimonies of India's alignment with the overarching objective of the Paris Agreement (Table 1.1).

Even if the goals formulated in 2015 were not particularly ambitious, they resulted in greater global visibility and provided the basis for better international assessment and comparison (Pillai & Dubash, 2021; Frank, 2016). India's "gift" of five more ambitious and advanced commitments to address the global climate challenge – domestically called *Panchamrit* – were approved by the Parliament in August 2022. Yet, according to the Climate Action Tracker project that assesses climate commitments of all signatories of the Paris Agreement, India's contributions – like that of many countries – remain highly insufficient to hold warming well below 2°C (CAT, 2023).

At the COP26 in Glasgow, Prime Minister Modi promoted the "one-word movement" for "LIFE" – a Lifestyle for Environment – advocating for environmentally conscious production, consumption, and supportive policy change. The India-led global movement was defined more closely by NITI Aayog and the Ministry of Environment, Forests, and Climate Change through a list of 75 actions across seven verticals from energy and water savings, the reduction of waste, single-use plastic and e-waste, a healthy lifestyle, and sustainable food systems. Examples brought forward by NITI Aayog feature plant-based diets and biodegradable utensils as well as adaptive architectural forms that naturally lower electricity consumption. Adopted by one billion people, this would have the potential to reduce GHG emissions by 20 per cent (NITI Aayog, 2022). A modelling study by the IEA (2023) that takes up India's LIFE mission highlights behavioural changes and sustainable consumer choices along the adoption of more energy-efficient household appliances, electric vehicles, and residential solar rooftop installations. The IEA concludes that consumers could save US$440 billion annually in energy bills and two billion tonnes of global carbon dioxide (CO_2) in 2030 if LIFE measures would be fully adopted globally. Moreover, India's role in the International Solar Alliance,[1] the Coalition for Disaster Resilient Infrastructure, and the Leadership Group for Industry Transition further corroborates the country's symbolic leadership as a "Middle Power" (Dubash, 2021) and a confident voice of the Global South. However, when we look at domestic conditions, which this book is fully dedicated to, the picture becomes more layered and nuanced.

Pillai and Dubash (2021) have traced the institutional history of climate politics in India since the early 1990s. They argue that today's climate policy lacks high-level strategic oversight, would be inconsistent and sectorally fragmented while a "thinly institutionalised co-benefits approach" to global warming dominates the growth and development paradigms. Yet no matter how contested domestic climate mitigation and adaptation efforts are, they build upon a frame that will be stabilising over time because the effects of climate change are already felt and are highly likely to only exacerbate both within India and beyond. India's net-zero pledge is an acknowledgement of this threat and shows its enlightened self-interest in being an integral part to the holistic transformation of how people live on this planet. By contrast, proposing a target of net-zero by 2070 also shows that India demands more time for a more just energy transition while both Western, industrial countries and rapidly growing emerging economies would have to be equally ambitious to address the climate challenge. Every new household or other entity connected to the Indian power grid and off-grid solutions should be considered an opportunity, if not outright necessity, to clean up the existing and expanding energy system.

At the Fourth Energy Forum in October 2020, Prime Minister Modi identified seven key drivers of India's energy map, which add an important layer to the internationally announced climate targets (ToI, 2020). These overarching directives are as follows:

- Accelerating efforts to move towards a gas-based economy.
- Cleaner use of fossil fuels like petroleum and coal.
- Greater reliance on domestic sources to drive biofuels.
- Achieving the renewable targets of 450 GW by 2030.[2]
- Increasing contribution of electricity to decarbonise mobility.
- Moving into emerging fuels, including hydrogen.
- Digital innovation across all energy systems.

This book reflects on developments in these different areas of India's emerging energy landscape. We start with a chapter on Just Transitions in India's coal sector. Coal is a natural reserve of India. Domestic coal production has been steadily increasing and is projected to grow further. Yet a global and domestic net-zero target brings fossil fuels and a Just Transition away from the most polluting energy carriers into focus.

Global climate justice and more poignantly a just energy transition are subjects of international debate for multiple decades. Indian civil society organisations have continuously been at the forefront of criticising the (emerging) international framework conventions from Agarwal and Narain's (1991) landmark paper on *Global Warming in an Unequal World* to Tongia and Kanitkar's analyses of climate justice aspects towards a carbon net-zero world (Pai & Jai, 2023). In over 30 years, the argumentative bottom line remains the same: climate conventions would inhibit the right to development by curtailing the use of fossil fuels while the "historical responsibility of the developed world to creating the problem of emissions is erased, and worse, the burden of future transition moves to the still developing world" (Narain, 2017). Kanitkar points at least-cost optimisation models for global carbon removal used in Intergovernmental Panel on Climate Change assessment reports, which put the mitigation burden to where it is the cheapest to lower emissions. Accordingly, if afforestation in the Global South and eliminating coal in Asia is cheaper than cutting oil and gas use in North America and Europe, these models would suggest exactly that without considering absolute and relative costs or developmental objectives such as zero hunger, malnutrition, universal housing, and employment for all (Pai & Jai, 2023).

Another case in point is energy services. Energy consumption in parts of the Global South is still significantly below that of the developed world. Yet speaking of an Indian Energy Revolution as the book title suggests might be most appropriate in the context of providing electricity access. Based on National Health Survey data, electricity access in India's poorest states made big leaps. In just five years, beginning with the first Modi administration in 2014, electricity access in the state of Bihar has risen from 15 to 83 per cent, from 19 to 63 per cent in neighbouring Uttar Pradesh, and from 27 to 69 per cent in Assam. In 19 out of 36 states and Union Territories, electricity access is now at above 95 per cent (Ravi, 2023). "Growing energy services will mean more carbon emissions . . . for some time," as Tongia points out (Pai & Jai, 2023). Given the increasingly limited carbon space available, industrially advanced, rich nations with high deposits of GHG emissions "must reduce, so we can grow" (Narain, 2017). An emissions rights framework is mathematically no issue even with conflicts around the recognition of historic emissions. However, implementing such framework even for the limited time frame of each NDC cycle has proved completely impossible. Based on the first set of NDCs submitted by China, the United States, and the EU, their emissions generated during 2011–2030 would be 218 Gt, 96 Gt, and 68 Gt of CO_2-eq, respectively. To stay within 1.5°C until 2100, the world emissions budget is 400 Gt at 66 per cent probability (IPCC, 2018). Accordingly, under current policies, China, the United States, and EU will have used up more than 95 per cent of the globe's carbon space already by 2030 (Ghosh &

Chaturvedi, 2015). At the same time, as illustrated by electricity access data, India won't wait for countries of the Global North to provide a fair share of the remaining carbon space. Another case in point is India's government seizing economic opportunity out of the European energy crisis resulting from Russia's military aggression on Ukraine. "If oil is available and at a discount, why shouldn't I buy it? I need it for my people," says Nirmala Sitharaman, India's finance minister (Schmall & Reed, 2022). Gatik Ship management – a corporation emerging out of nothing – has been instrumental in facilitating one of the most significant shifts in global oil flows in decades. Prior to Russia's invasion of Ukraine in February 2022, the oil giant supplied less than 1 per cent of India's crude. It now accounts for around 30 per cent (FT, 2023). Even if it might be too early to deeply look into the medium- and long-term effects of the reshuffling of the global energy landscape, *Gatik* in Sanskrit means "speed" and India will be swift in feeding its energy hunger whether that is conforming with Paris Agreement targets or not. Meanwhile, the climate justice debate continues around carbon pricing, also of historic emissions, and technology transfers (Pai & Jai, 2023).

One driver of India's energy future will be the cleaner use of petroleum derivatives and coal. Coal electrification has more than doubled over the past decade and might keep rising over the long term to provide a reliable baseload for India's economic growth. Yet, with four times more people, India's coal plant capacity is at par with that of the United States (11 per cent) and with same population size significantly behind that of China (51 per cent). Supercritical (high pressure) coal plants can be up to one-third more efficient than subcritical (low-pressure) coal plants and generate correspondingly less CO_2 (Frank, 2016). However, in a recent announcement, an amendment to the final draft of the National Electricity Policy states that India will pause plans to add new coal power plants, if approved by the Parliament (Reuters, 2023).

Chapter 2 of this book deals with India's inevitable coal-phase down. Chandra Bhushan and Srestha Banerjee, two progressive environmental thinkers and doers, deliberate upon governance mechanisms and policy approaches to a more equitable energy transition in India and the Global South more broadly. Based on in-depth research in India's coal mining regions, they describe the local realities and develop a holistic framework for Just Transition planning in India's coal states.

Over the twentieth century, oil and natural gas have pushed coal use aside in global primary energy consumption (Ritchie, 2019). For Indian stakeholders, the historic example of the United States and United Kingdom are still relevant cases for building a gas-based economy even when natural gas – a fossil energy carrier – is considered only a bridge fuel and technology towards a low-carbon energy system. Two recent European publications break with that paradigm. With a report on the "Future of Gas," the Science Advisory Council of the European Academies has stepped forward declaring that "[n] atural gas is not cleaner than other fossil fuels and using it instead of coal or oil risks achieving little or no reduction in greenhouse gas effects" (EASAC, 2023). With *Breaking Free from Fossil Gas. A New Path to a Climate-Neutral Europe*, Agora Energiewende (2023) – a Berlin-headquartered think tank has presented a landmark study that lays out Europe's move away from even the cleanest fossil fuel. The burden of transitioning into a clean energy future, as Narain (2017) would put it, does not move to the still developing world alone. We all know that the issue of climate change leaves no option. It is a make-or-break situation for us all while the build-up of India's energy system is at a similarly critical juncture.

The International Monetary Fund estimates the Indian economy to grow at around 6 per cent in 2023 and 2024 (IMF, 2023). Indian policymakers aspire to a US$5 trillion economy at around 2025 and a US$10 trillion economy in the 2030s, with India becoming the third-largest economy globally. Significantly more energy is required to drive this growth. When Prime Minister Modi presented seven key drivers of India's energy map – the first driver listed was the acceleration of efforts to move towards a gas-based economy corroborated by plans to increase the share of natural gas in the primary energy mix from current 6 per cent to 15 per cent by 2030.

In Chapter 3, we will look at how global industrial development was fuelled by conventional fossil sources of energy for more than 200 years and how India's economic development until the turn of the millennium was no different. Even if the Indian economy is far from being as developed as that of Western nations or China to the East, the country is confronted with a similarly profound transformation of its existing energy system. In fact, in the Indian mindset, the legacy of a traditional development paradigm is lingering and despite the enormous potential to develop a more sustainable energy system, in spite of the risk of a carbon lock-in, stranded assets, and potential energy insecurity, natural gas is a prime fuel of choice. In this chapter, Akhil Mehrotra, Aayushi Agarwal, and I present a road map for a cleaner and more prosperous India by critically reflecting on the various challenges constraining the gas industry.

One step ahead, another key driver in energy landscape development recognised and promoted by the Indian Government is in emerging fuels such as hydrogen. When, in August 2021, India's National Hydrogen Mission was published, this showed that "unlike solar and battery technologies, Indian policy action in Green Hydrogen has not lagged the global momentum" (Amitabh Kant, G20 sherpa and former CEO of NITI Aayog). In 2022, NITI Aayog presented a report outlining five key action requirements to *Harnessing Green Hydrogen* informing a National Action Plan in the making (NITI Aayog and RMI, 2022). In Chapter 4, on "Green Hydrogen: India's Billion-Dollar Opportunity," Tirtha Biswas and Deepak Yadav from the Council on Energy, Environment and Water fathom the country's potential to produce globally affordable green hydrogen at large scale for both domestic purposes and export markets. They share their technical insights, present calculations, and show a viable way forward in decarbonising hard-to-abate sectors of the Indian economy. However, as Chaturvedi and Malyan (2022) point out, developing hydrogen is critical but no panacea to the country's net-zero challenge. The second "break-through" technology that Chaturvedi and Malyan consider is Carbon Capture and Storage (CCS), which would allow for significantly higher consumption of crude oil, significantly lower the use of biofuels, a delayed phase-out of coal, and a consequently lower installed capacity of solar energy in the Indian economy. India's first test site for carbon capture, utilisation, and storage was inaugurated at National Thermal Power Corporation (NTPC)-run 500 MW Vindhyachal Super Thermal Power Station in Madhya Pradesh in August 2022.

Given that India is both the second-largest producer and consumer of cement as well as the second-largest crude steel producer globally (IBEF, 2022; World Steel Association), Paltsev et al. see demand surge such as through road construction, urban infrastructure, and commercial real estate to negate the emissions reduction potential of electrification, natural gas use, and resource efficiency measures. The authors project CO_2 emissions from hard-to-abate sectors to rise by about 2.6 times between 2020 and 2050, pointing at the need for carbon pricing to make CCS technology economically viable (Paltsev et al., 2022). A chapter analysing the discourse around carbon pricing was planned but fell victim to the pandemic. In addition, a more substantial debate seems to arise in India only now triggered by the European Carbon Border Adjustment Mechanism.

Pillai and Dubash (2021) question the government's capability of restructuring incentives which would mean to increase costs for all parties involved. Yet, learning and investment cycles, the charging of a CO_2 price for internationally traded goods, the 50-plus-year lifetime of industrial sites and other infrastructure suggest that Indian stakeholders start exploring net-zero carbon technology and finance solutions early.

What we know is that India's first set of NDCs demanded US$2.5 trillion (2014–2015 prices) of investment. Moody's reckons with a need of US$225–250 billion to meet India's 280 GW of solar and 140 GW of wind capacity installation. While Asia-Pacific countries hold four in ten places of energy transition investment, India and South Korea joining China and Japan (BNEF, 2022), there is no transparent and complete overview of who invests what, when, how, and to what effect into India's energy transition.

India today has the largest concentration of young people in the world ever recorded in history. More than 420 million Indians – full of aspiration – are between 15 and 34 years of age. For the next several years, one million Indians are turning 18 every single month (Sengupta, 2016). According to Nagarajan (2016), we will be seeing India's population growth peaking perhaps at 1.75 billion only in 2060. Thus, to allow a large part of the Indian population to live up to its (basic) needs is not only an issue of domestic intergenerational justice but distributive justice more broadly. India today is also the most populous country in the world, home to 18 per cent of the global population. Emissions generated in the country amounted to 3.4

Gt in 2019 (CAIT database) constituting 6.9 per cent of global total GHG emissions and resulting in 2.4 tonnes of CO_2-eq per capita. This is still markedly below the global average of 6.3 tonnes and correlates with a Human Development Index rating classified as of very low quality of life. The country's aggregate and per capita GHG emissions will keep rising over the next decades. However, the foundations for the systemic transformation towards cleaner and more efficient energy technologies have been laid and already show substantive results. In November 2021, non-fossil power capacity has reached 40 per cent – one goal in India's first set of NDCs to mitigate climate change. This achievement includes a more than fivefold increase in installed capacity – not generating capacity – of new renewable sources of energy to 105 GW covering wind, solar, bio, and small hydropower (CEA, 2011, 2021; PIB, 2021). This also includes a small fraction of nuclear power (6.8 GW). With the revised set of NDCs presented at the COP26 in Glasgow, including a target of 50 per cent non-fossil fuel energy capacity by 2030, the question of the importance of nuclear power in India's future energy mix got into focus.

Nuclear energy is generally considered a clean source of power production that has the advantage of providing continuous base load over wind and solar energies. While global electricity generation from nuclear power plants has dropped below 10 per cent in 2021 for the first time in four decades, China is leading in expanding nuclear power generation (WNISR, 2023). In a recent article, prominent economic and energy experts of India speak up for higher shares of nuclear energy in the country's power mix and make policy reform suggestions to eliminate the bottlenecks of India's nuclear energy programme (Ahluwalia et al., 2023). In Chapter 5, Dr A.T. Ajmal Khan critically reflects on the future of nuclear energy in India's sustainable development pathway drawing from the past and present of stakeholders and respective institutional positioning in India's nuclear energy sector.

At the end of 2021, the installed capacity of the Indian power sector was 393.4 GW, roughly six times the capacity of the year 1990. With that, electricity generation capacity has increased annually at 12 per cent on average. Data collected by the Central Electricity Authority show that all sources contribute to this growth while its composition is relatively dynamic over the years. In 2002, for example, coal-based capacity stood at 59 per cent (62.1 GW) and new renewable energy sources at merely 2 per cent (1.6 GW). In 2021, the share of coal in installed capacity has come down to 53 per cent (209.8 GW) and the share of new renewables increased to 26.7 per cent (104.9 GW) (NPP, 2022). Despite the massive growth in new and renewable energy, including large hydropower to 151.4 GW (38.5 per cent), both the near-term target of 175 GW by 2022 and the 2030 target seem out of reach. If India wants to achieve 450 GW if not 500 GW capacity from renewable energy sources by 2030 – the fourth goal on India's energy map and one of the revised NDC targets – unprecedented new installations of up to 38.7 GW annually would be needed.

Both the biopower and hydropower sectors have (almost) achieved their 2022 expansion targets. Solar and wind power installations, by contrast, show a significant backlog. The shortcoming is most prevalent in the solar sector, where only half of the expansion target has materialised thus far. In Chapter 6, Mr Parvej analyses the solar power sector by delving into the socio-economic and technical complexities that his research on perceptions on solar photovoltaic installations revealed. Based on the results of his local study in the state of Haryana, North India, Parvej argues for an integrated capacity building and participatory approach to address the various concerns raised by different actor groups to achieve enhanced generation and utilisation of solar energy in India consistent with self-imposed targets. In the subsequent Chapter 7, Deepak Sangroya, Kishore Kumar Gangwani, and Sanjay Chaudhury share their insights into the conditions of the Indian wind energy sector, which in 2022 was 20 GW behind the installation target. They find that India currently taps only 4 per cent of its wind energy potential and identify several key barriers to wind energy adoption. The authors attach great importance to financial means to overcome these obstacles and call for government intervention to improve the financial viability of distribution companies.

In Chapter 8, Rishu Garg and her colleagues from Centre for Study of Science, Technology and Policy help us better understand the institutional challenges to a cleaner grid revolving around India's distribution companies. Here, the seventh driving force of India's energy future – digital innovation across all energy systems – is explicitly addressed first. However, a separate chapter on digital innovations in India's energy system such as smart meters, artificial intelligence, machine learning, digital ledger technology, and internet of things did not materialise due to a lack of substantial technological development beyond the level of pilot and demonstration projects. By contrast, even though India is nowhere near the economically optimal point of integrating renewable energies at which grid-scale energy storage would be needed, we have included a chapter on "Energy Storage – Enabling a Greener Grid" (Chapter 9). Here Rahul Walawalkar and Debmalya Sen present several case studies at various levels of the energy storage value chain to discuss different applications that energy storage technologies find in an energy system that is expected to witness a major growth of renewable sources of energy.

In Chapter 10, Abhijit Banerjee of O.P. Jindal Global University challenges the government's belief in greater reliance on domestic sources to spur biofuels – the fifth driving force of India's energy scenario – by presenting "A Realistic Assessment of the Prospects of Biofuels for Transportation in India." He argues that the Government of India would be better advised by pursuing a different strategy for the development of a more sustainable transportation sector. It is the sixth driving force named by Prime Minister Modi that speaks of increasing contributions of electricity to decarbonise the transport sector. Even if only 9 per cent of the total national CO_2-eq emissions are generated in the Indian transport sector compared to roughly 25 per cent globally, the sales statistics show a clearly emerging trend in demand for more and bigger internal combustion engine (ICE) vehicles (Ge et al., 2020; Miller & Façanha, 2014; IPCC, 2014; SIAM, 2021). Thus, without the shift to low-carbon vehicle technologies, India risks making similar development mistakes as Western nations. To set the course right, the Government of India launched the National Electric Mobility Mission Plan in 2013 supported through the Faster Adoption and Manufacturing of Electric Vehicle programme. In Chapter 11, I present an ecosystem perspective broaching multiple issues through "Business Model Innovations for Electric Vehicle Adoption in India."

Electric vehicles combine multiple benefits for a developing economy. Battery electric vehicles (BEVs) don't contribute to local air and sound pollution and don't generate any direct atmospheric pollution (CO_2) when powered by renewable, non-fossil sources of energy. Even if the mineral resource availability (lithium, graphite, nickel, cobalt, etc.) for BEV production in India is creating new challenges, the significantly lower number of parts needed in a BEV (around 200 components), compared to ICE vehicles (roughly 1,400 parts) and the higher share of electric and electronic parts (FES, 2010; Fraunhofer ISI, 2012), might lower the entry threshold for BEV production and job creation in an emerging economy like India. Moreover, when comparing the energy efficiency of a BEV to that of a petrol- or diesel-fuelled vehicle, the BEV converts 86 per cent of electricity directly into motion, whereas the fossil-run engines reach around 30 per cent efficiency at best (BMU, 2013).

Energy efficiency is the elephant in the room. Energy efficiency is a highly potent solution to energy dilemmas that also the Prime Minister turns a blind eye to when presenting his seven key drivers of India's energy map. Energy efficiency measures, more than renewable energies and nuclear energy together, have the potential to reduce GHG emissions without compromising growth and development (IEA, 2021). Different from many countries, India's rural population still outweighs urban dwellers by a ratio of roughly 7:3. Over the next two decades, India's urban population will increase by about 270 million people due to both population growth and various urban pull factors. The housing needs and related infrastructure requirements will not only spur demand in construction material but also provide the opportunity to build more efficiently. The Energy Conservation Building Code is currently and incomprehensibly only applicable to large commercial buildings, the UJALA LED programme, the star rating for (home) appliances, and the mandatory energy-saving targets for energy-intensive industries under the Perform, Achieve, and Trade scheme are important components of India's National Mission for Enhanced Energy Efficiency. However, much more can and must be achieved. We close this edited volume with a contribution by Ramya Natarajan, Kaveri Ashok, and Sarah Khan dedicated to India's urbanisation. In "Systems Thinking for Doughnut Cities," they take a holistic approach applying Kate Raworth's model of a sustainable society to India's urbanisation process.

In 2022, India celebrated 75 years of independence and has great plans for its anniversary in the year 2047. India and its people have already come a long way in regaining confidence and control over their matters, culminating in the wish to become more self-reliant. Yet, to secure our future requires a "one-world movement" in which growing national strength and decisive action build on ecologically justifiable economic concepts and nature conservation as well as social justice, equality, and international collaboration for the greater good of all.

Notes

1 At ISA, India has proposed the "One Sun, One World, One Grid" initiative – an equivalent to China's Belt and Road Initiative, though with less geopolitical ambition.

2 The revised set of NDCs issued in 2021 has increased ambition to 500 GW renewable energies by 2030.

References

Agarwal, A. & Narain, S. (1991). *Global Warming in an Unequal World. A Case of Environmental Colonialism*. Centre for Science and Environment, New Delhi.

Agora Energiewende (2023). *Breaking Free from Fossil Gas. A New Path to a Climate-Neutral Europe*. May 2023. Retrieved from: https://static.agora-energiewende.de/fileadmin/Projekte/2021/2021_07_EU_GEXIT/A-EW_292_Breaking_free_WEB.pdf (Last accessed: 09.06.2023)

Ahluwalia, M.S., Patel, U. & Tongia, R. (2023). Does nuclear power have a role in our climate change strategy? *Live Mint*. 04.05.2023. Retrieved from: www.livemint.com/opinion/columns/expanding-nuclear-power-in-india-opportunities-and-challenges-for-private-sector-participation-and-policy-reform-11683139229040.html (Last accessed: 12.05.2023)

BMU (2013). *Erneuerbar mobil. Marktfähige Lösungen für eine klimafreundliche Elektromobilität (Renewable Mobile. Marketable Solutions for Climate-Friendly Electric Mobility)*. Bundesministerium für Umwelt, Berlin, April 2013.

BNEF (2022). *Global Investment in Low-Carbon Energy Transition Hit $ 755 Billion in 2021*. Bloomberg New Energy Finance. 27.01.2022. Retrieved from: https://about.bnef.com/blog/global-investment-in-low-carbon-energy-transition-hit-755-billion-in-2021/ (Last accessed: 21.06.2023)

CAT (2023). *Climate Action Tracker by Country*. Last update June 2023. Retrieved from: https://climateactiontracker.org/countries/# (Last accessed: 27.06.2023)

CEA (2011). *Annual Report 2010–11*. Central Electricity Authority. June 2011. Retrieved from: https://cea.nic.in/wp-content/uploads/2020/03/annual_report-2011.pdf (Last accessed: 21.05.2021)

CEA (2021). *Executive Summary on Power Sector*. Central Electricity Authority. March 2021. Retrieved from: https://cea.nic.in/wp-content/uploads/executive/2021/03/executive.pdf (Last accessed: 21.05.2021)

Chaturvedi, V. & Malyan, A. (2022). Implications of a net-zero target for India's sectoral energy transitions and climate policy. *Oxford Open Climate Change*, 2(1): kgac001. Retrieved from: https://doi.org/10.1093/oxfclm/kgac001 (Last accessed: 15.06.2023)

Dubash, N.K. (2021). *Whither International Leadership in Climate Diplomacy: An Ambition Check on the Paris Agreement and International Mitigation Efforts.*

Climate Change Mitigation Webinar Series. Center for Environmental Policy, School of Public Affair, American University, Washington, DC.

EASAC (2023). *Future of Gas*. European Academies' Science Advisory Council. 24.05.2023. Retrieved from: https://easac.eu/publications/details/future-of-gas (Last accessed: 09.06.2023)

FES (2010). *Zukunft der deutschen Automobilindustrie. Herausforderungen und Perspektiven für den Strukturwandel im Automobilsektor (Future of the German Automotive Industry. Challenges and Perspectives for Structural Change in the Automotive Sector)*. WISO Diskurs, Friedrich-Ebert-Stiftung, Berlin. December 2010.

Figueres, C. et al. (2017). Three years to safeguard our climate. *Nature*, 546: 593–595. 29.06.2017.

Frank, C. (2016). *India: Potential for Even Greater Emissions Reductions*. Retrieved from: www.brookings.edu/wp-content/uploads/2016/07/india-frank.pdf (Last accessed: 03.06.2021)

Fraunhofer ISI (2012). *Technologie-Roadmap Energiespeicher für die Elektromobilität 2030 (Technology Roadmap Energy Storage for Electric Mobility 2030)*. Fraunhofer-Institut für System- und Innovationsforschung (ISI), Karlsruhe.

FT (2023). The unknown Indian company shipping millions of barrels of Russian oil. *Financial Times*. 04.05.2023.

Ge, M., Friedrich, J. & Vigna, L. (2020). *4 Charts Explain Greenhouse Gas Emissions by Countries and Sectors*. World Resources Institute, Washington, DC. Retrieved from: www.wri.org/insights/4-charts-explain-greenhouse-gas-emissions-countries-and-sectors (Last accessed: 21.02.2022)

Ghosh, A. & Chaturvedi, V. (2015). For a few degrees more. *Business Standard*, XXII(176). 22.12.2015.

HT (2021). PM Modi delivers India 'panchamrit' gift at COP26 to fight climate change: Five commitments in detail here. *Hindustan Times*. 01.11.2021. Retrieved from: www.hindustantimes.com/india-news/pm-modi-delivers-india-panchamrit-gift-at-cop26-to-fight-climate-change-five-commitments-in-detail-here-101635788023755.html (Last accessed: 22.02.2022)

IBEF (2022). *Indian Cement Industry Analysis*. Indian Brand Equity Foundation. Retrieved from: www.ibef.org/industry/cement-presentation (Last accessed: 21.06.2023)

IEA (2021). *India Energy Outlook 2021*. International Energy Agency (IEA), Paris. February 2021. Retrieved from: www.iea.org/reports/india-energy-outlook-2021 (Last accessed: 12.03.2021).

IEA (2023). *LiFE Lessons from India*. International Energy Agency (IEA), Paris. Retrieved from: https://www.iea.org/reports/life-lessons-from-india (Last accessed: 18.01.2024)

IMF (2023). *World Economic Outlook Projections*. International Monetary Fund. April 2023. Retrieved from: www.imf.org/en/Publications/WEO/Issues/2023/04/11/world-economic-outlook-april-2023?s=03 (Last accessed: 02.06.2023)

IPCC (2014). *Summary for Policymakers. Climate Change 2014. Mitigation of Climate Change. Contribution of Working Group III to the Fifth Assessment Report of the Intergovernmental Panel on Climate Change*. Cambridge University Press, Cambridge, UK and New York, USA.

IPCC (2018). Summary for policymakers. In: Masson-Delmotte, V., Zhai, P., Pörtner, H.-O., Roberts, D., Skea, J., Shukla, P.R., Pirani, A., Moufouma-Okia, W., Péan, C., Pidcock, R., Connors, S., Matthews, J.B.R., Chen, Y., Zhou, X., Gomis, M.I., Lonnoy, E., Maycock, T., Tignor, M. & Waterfield, T. (Eds.), *Global Warming of 1.5°C: An IPCC Special Report on the Impacts of Global Warming of 1.5°C above Pre-Industrial Levels and Related Global Greenhouse Gas Emission*

Pathways, in the Context of Strengthening the Global Response to the Threat of Climate Change, Sustainable Development, and Efforts to Eradicate Poverty. Cambridge University Press, Cambridge and New York, NY, pp. 3–24. https://doi.org/10.1017/9781009157940.001

Miller, J.D. & Façanha, C. (2014). *The State of Clean Transport Policy. A 2014 Synthesis of Vehicle and Fuel Policy Developments*. ICCT, Washington, DC.

Nagarajan, R. (2016). Nein, es gibt keine Bevölkerungsexplosion in Indien (No, there is no population explosion in India). In: *Der Spiegel*. Der Spiegel, Hamburg.

Narain, S. (2017). *Conflicts of Interest. My Journey through India's Green Movement*. Viking by Penguin Random House India, Gurgaon.

NITI Aayog and RMI (2022). *Harnessing Green Hydrogen. Opportunities for Deep Decarbonisation in India. Report*. June 2022. Retrieved from: www.niti.gov.in/sites/default/files/2022-06/Harnessing_Green_Hydrogen_V21_DIGITAL_29062022.pdf (Last accessed: 15.06.2023)

NPP (2022). *All India Installed Capacity of Power Stations*. National Power Portal. Retrieved from: https://npp.gov.in/public-reports/cea/monthly/installcap/2021/DEC/capacity1-2021-12.pdf (Last accessed: 04.02.2022)

Pai, S. & Jai, S. (host) (2023). Path to climate justice in a net zero world. Ft. Tejal Kanitkar & Rahul Tongia. *The India Energy Hour*, audio podcast episode 3. 16.05.2023. Retrieved from: https://theindiaenergyhour.com/episode/path-to-climate-justice-in-a-net-zero-world/?s=03 (Last accessed: 30.05.2023)

Paltsev, S. et al. (2022). Economic analysis of the hard-to-abate sectors in India. *Energy Economics*, 112(2022): 106149.

PIB (2021). *India has Achieved Its NDC Target with Total Non-Fossil Based Installed Energy Capacity of 157.32 GW which is 40.1% of the Total Installed Electricity Capacity*. Press Information Bureau Government of India. Retrieved from: www.pib.gov.in/PressReleasePage.aspx?PRID=1785808 (Last accessed: 25.01.2022)

Pillai, D.V. & Dubash, N.K. (2021). The limits of opportunism: The uneven emergence of climate institutions in India. *Environmental Politics*, 30: 93–117. https://doi.org/10.1080/09644016.2021.1933800

Ravi, S. (@ShamikaRavi) (2023). India's real economy underwent a revolution with mere 5 years. Access to electricity among the poorest population went up significantly [Tweet]. Twitter. 16.01.2023. Retrieved from: https://twitter.com/ShamikaRavi/status/1616676973677539329?t=MeQcW3saIKfICFpFJWNXEA&s=03 (Last accessed: 01.06.2023)

Reuters (2023). *India Amends Power Policy Draft to Halt New Coal-Fired Capacity*. 05.05.2023. Retrieved from: www.reuters.com/business/energy/india-amends-power-policy-draft-halt-new-coal-fired-capacity-sources-2023-05-04/ (Last accessed: 07.06.2023)

Ritchie, H. (2019). *Global Primary Energy Consumption by Source*. Our World in Data. Retrieved from: https://ourworldindata.org/energy-mix (Last accessed: 22.04.2021)

Schmall, E. & Reed, S. (2022). India finds Russian oil an irresistible deal, no matter the diplomatic pressure. *New York Times*. 04.05.2022.

Sengupta, S. (2016). Every month for the next several years, 1 million Indians will turn 18. *The Guardian*. 24.04.2016.

SIAM (2021). *Automobile Domestic Sales Trends*. Society of Indian Manufacturers. Retrieved from: www.siam.in/statistics.aspx?mpgid=8&pgidtrail=14 (Last accessed: 20.02.2022)

ToI (2020). PM Modi lays down India's energy map, list seven key drivers of change. *Times of India*. 26.10.2020. Retrieved from: https://timesofindia.indiatimes.com/business/india-business/pm-modi-lays-down-indias-energy-map-lists-seven-key-drivers-of-change/articleshow/78874616.cms (Last accessed: 18.04.2021)

UNFCCC (2015). *India's Intended Nationally Determined Contributions: Working towards Climate Justice*. United Nations Framework Convention on Climate Change, Bonn. Retrieved from: https://www4.unfccc.int/sites/submissions/INDC/Published%20Documents/India/1/INDIA%20INDC%20TO%20UNFCCC.pdf (Last accessed: 08.07.2020).

WNISR (2023). *The World Nuclear Industry Status Report 2022*. Retrieved from: www.worldnuclearreport.org/IMG/pdf/wnisr2022-v3-lr.pdf (Last accessed: 12.05.2023)

2 A Just Energy Transition

Considerations for India's Coal Sector

Chandra Bhushan and Srestha Banerjee

2.1 Introduction

Over the past two decades, the urgency to build a low-carbon society has gained momentum from policy corridors to broader public discourse. The growing body of research on climate change has made it clear that a shift away from the fossil-fuel economy is inevitable. The questions are only by when and how. The 2015 Paris Agreement and the special report of the Intergovernmental Panel on Climate Change (IPCC) in 2018 have set the time frame for the transition. To avoid the catastrophic impacts of climate change, the world needs to limit global warming well below 2°C above pre-industrial levels and pursue efforts to limit the temperature increase even further to 1.5°C (UNFCCC, 2015). For this, global carbon dioxide (CO_2) emissions will have to be reduced by half by 2030 from 2010 levels and reach net-zero by 2050 (UN, 2020). This will necessitate a massive reduction in the use of coal, as it accounts for nearly 44 per cent of global CO_2-equivalent emissions (IEA, 2018). Therefore, the IPCC has recommended the phasing out of coal-based power plants by 2050 and adopting a "system transition" in electricity generation, largely relying on renewable energy (RE) sources (IPCC, 2018). Oil and natural gas will also need to be phased out over time, but coal is seen as a priority internationally, mainly because affordable and reliable alternatives already exist for the electricity sector.

Coal-based electricity has already started to face headwinds. A mix of environmental and energy policies and competition from cost-efficient renewables is steadily squeezing coal out of the global electricity mix. In fact, the contribution of coal to global electricity generation has reduced from 40 per cent in 2010 to 36.4 per cent in 2019 (BP, 2020). Global investments in new coal-fired power projects have also been declining over recent years (IEA, 2019). Both the US Energy Information Administration (USEIA, 2018) and the International Energy Agency (IEA, 2016) have predicted a significant decline in coal consumption in the years ahead. In India, while the total coal consumption remains robust, coal's competitive edge in the power sector is gradually declining, with implications for both the coal mining and coal-based power sectors.

DOI: 10.4324/9781003281818-2

Considering the various environmental, social, and economic factors and the imperative to avoid the unexpected closure of coal mines, it is time we begin planning a transition away from a coal-based economy in a deliberate and phased manner. However, such a shift will not be a simple one. It will be a complex and long-term process involving all stakeholders and governments at all levels.

2.2 What Is a Just Transition and Why Is It Important for India?

The socio-economic aspects of the energy transition process have become an essential component of the climate change discourse. Building on demands of labour rights and environmental justice movements, there is a growing recognition that workers and local communities should not be the only stakeholders shouldering the burden of a transition away from fossil fuels. The basic idea of a Just Transition, therefore, is to ensure decent work opportunities and a social support system for the people whose livelihood is going to be affected by the transition. Simultaneously, the economic restructuring towards a low-carbon future that guarantees sustainable growth needs to be planned, engaging various stakeholders in the decision-making process.

The discussion on Just Transition so far has mostly been limited to countries of the Global North (Pai et al., 2020). The conversation has not taken roots in major coal economies of the Global South, such as China, India, and Indonesia. However, considering the collective response required for climate action, a discussion on planning a transition away from the coal economy should begin in these countries as well. For India, ambitious global climate action is necessary, considering the looming economic and social costs of the associated crisis. As per the Global Climate Change Risk Index 2020, India is the fifth most vulnerable country to climate change impacts (Eckstein et al., 2020). Studies suggest that under a business-as-usual scenario the country risks losing 2.5 to 3 per cent of its gross domestic product (GDP) by 2050 to the climate crisis (Mani et al., 2018). It is in India's interest that the world meets the 1.5°C goals; for this, India will have to take a leadership role in climate action. By putting a Just Transition at the centre of its climate mitigation plan, India can send a strong leadership signal to the world. Just Transition will not only help India push for global collective action, but it will also help in building a sustainable and resilient economy in the country's coal-dependent regions.

2.3 Objective, Scope, and Approach of the Chapter

This chapter looks into the question of Just Transition in India in the context of the socio-economic realities of the coal sector. A transition away from petroleum and natural gas is expected to have a much longer time horizon and will therefore remain outside the scope of this chapter. Since any approach to Just Transition needs to be evaluated in a country- and region-specific manner, the chapter will particularly dwell on two important aspects. First, what does justice mean for coal-dependent communities in India, and, second, what policy approach and governance mechanisms need to be considered to enable a process of Just Transition for these regions. The research and analysis are based on literature review, analysis of government data and policies as much as ground observations by the author undertaken in key coal districts during 2020–2021.

The following section (Section 2.4) provides an overview of Just Transition in the international context. Section 2.5 outlines the background of the coal sector in India and its socio-economic importance. Section 6 analyses the rationale for Just Transition in the Indian coal sector. Section 7 presents a broad framework for Just Transition planning and policies for the Indian coal sector before overall conclusions are drawn in Section 8.

2.4 Just Transitions in the International Context

The inception of the Just Transition concept can be traced back to the 1990s when American labour unions demanded support for workers displaced by environmental protection policies (Morena et al., 2019). Given its roots, some of the most prominent framings of Just Transition come from labour unions, with a strong underpinning of "labour environmentalism" (Stevis et al., 2018). For example, the International Trade Union Confederation (ITUC) notes that Just Transition is "an economy-wide process that produces the plans, policies, and investments that lead to a future where all jobs are green and decent, emissions are at net zero, poverty is eradicated, and communities are thriving and resilient" (ITUC, 2017). However, over the years, a growing engagement of various stakeholder groups in the discourse has brought in transformative ideas. As has been very aptly described by the Climate Justice Alliance (2018), "Just Transition is a vision-led, unifying and place-based set of principles, processes, and practices that build economic and political power to shift from an extractive economy to a regenerative economy." The fundamental idea is that a healthy economy and a clean environment can and should coexist (just Transition Alliance, n.d.).

Just Transition, in all its framings, therefore, is both constructive and inclusive. With its recognition of the need to decarbonise the economy, while insisting that coal-industry workers and communities in coal-mining areas should not become the victims of the decarbonisation process, the concept of "Just Transitions" balances the obligations of climate action with social justice. It seeks to build an equitable and resilient economy leading to an improved quality of life for all. This philosophy is also reflected in the 2030 Agenda for Sustainable Development adopted by the United Nations (UN) General Assembly in October 2015. Several Sustainable Development Goals (SDGs), including SDGs 7, 8, 12, and 13, provide the impetus for a fair transition from fossil fuels, balancing the needs of secure and affordable energy, full and productive employment with the sustainable use and management of resources. The concept of Just Transition was brought onto the climate policy platform in 1997, with the ITUC including a demand for a Just Transition in its statement at the Kyoto Conference (Morena et al., 2019). Over the next two decades, the idea was highlighted continuously and advocated by various trade unions (particularly of the Global North) and found mention in key climate change deliberations. However, what has been missing since the Paris Agreement and the Katowice Conference is a clear inclusive pathway that operationalises Just Transition within the United Nations Framework Convention on Climate Change (UNFCCC). The Katowice Conference of Parties was considered an opportunity to build the international community's attention on the justice dimensions of climate mitigation and adaptation. But it exposed the gap between the understanding of climate policymakers of Just Transition and the complex and multifaceted ground reality that "lie deep in the everyday experiences of workers and frontline communities" (Morena et al., 2019). Therefore, much is left to be done at the UNFCCC to make the transition process inclusive and collective.

In the past few years, Just Transition has gained traction in the Global North, with several developed economies pledging to phase out coal over the next two decades and develop coal phase-out strategies, along with Just Transition plans. For this, policy initiatives and actions are being undertaken individually by governments as well as through alliances. With respect to country-level initiatives, some critical examples come from Canada, Germany, and Spain.

2.4.1 *Canada*

In 2016, Canada announced that it would phase out traditional coal-based power by 2030. In April 2018, the government also launched a "Task Force on Just Transition for Canadian Coal Power Workers and Communities" to better understand the impacts of phasing out coal and the support required for people directly affected (Government of Canada, 2018). Members of the task force visited four mining provinces, interacted with all potentially affected stakeholders, and developed recommendations for a sustainable transition away from coal. The task force highlighted certain key elements of Just Transition centred on sound policies and planning that take the local context into account and engage the local community in the process (Government of Canada, 2018).

2.4.2 *Germany*

Similarly, in 2018, Germany appointed a dedicated agency called "Commission on Growth, Structural Economic Change, and Employment" to recommend concrete measures and timelines that can enable the country's energy sector to achieve its emission reduction targets by 2030, as well as provide a blueprint for Just Transition of the coal workers and mining regions. The

Commission was entrusted to suggest job creation prospects, economic policy instruments, required investments in regions, and economic sectors support structural change. The Commission recommended a road map for coal phase-out by 2035 or latest by 2038, which was accepted by the German cabinet in January 2020 (The German Coal Commission, 2019). The coalition agreement of the new federal government moved the coal phase-out year to even 2030 (TS, 2021).

2.4.3 Spain

In 2018, Spain proposed a Just Transition deal that replaced coal industry subsidies with a sustainable development plan for the regions where coal mining closures were proposed. The deal – "Plan del Carbón" – was struck between the Spanish government and the mining unions, following a plan to close all economically unviable coal mines. The deal involved a EUR 250 million plan for five years (2019–2023) to train and retrain the people who lost their jobs from mine closures; provide for early retirement; sustainably restore former mining sites; recover forests; improve utility infrastructure; and upgrade facilities in the mining communities, including facilities for waste management, recycling, and water treatment (Just Transition Centre, 2019).

2.4.4 Other Initiatives

Just Transition is also being facilitated through several coalitions and alliances. In 2017, the United Kingdom and Canada founded the Powering Past Coal Alliance (PPCA) to accelerate the transition from traditional coal-fired electricity to clean energy. The target year for such transition has been set as 2030 for member countries of the Organization for Economic Cooperation and Development, and as 2050 for all countries. The alliance has members comprising national and sub-national governments, as well as businesses based in North America, South America, Europe, and to a lesser extent, in Africa (PPCA, 2020). The globally largest coal economies, including China, the United States, Australia, and India, have, however, shied away from joining this alliance.

The European Commission launched "The Coal Regions in Transition" initiative for its member states in 2018 with a similar objective as PPCA. At present, 20 coal regions spread across nine European Union countries are actively participating in the initiative (EC, 2020a). In January 2020, the European Commission further adopted a proposal of creating a Just Transition fund to facilitate this initiative. The fund will support workers and encourage companies and regional authorities to make investments that can enable a Just Transition (EC, 2020b).

However, the idea of Just Transition has not gained much prominence in the major coal-dependent countries of the Global South. The dichotomy of jobs versus the environment is still dominant in these economies due to their high reliance on coal. Further, the discourse on the right to development, growing energy demands, and arguments of a low share of greenhouse gas (GHG) emissions have obstructed the discussion around Just Transitions (UNRISD, 2018). However, 18 countries of the Global South are signatories of the Declaration on Just Transition adopted in Katowice in 2018. By contrast, two of the biggest coal economies, India and China, decided not to be part of that Declaration.[1]

2.5 The Indian Coal Sector

Coal has been intricately linked with India's economy ever since colonial times. Since independence, coal has been viewed as the backbone of India's industrial and energy security. India has roughly 170 billion tonnes of proven coal reserves, making it one of the top in the world (MoSPI, 2022). In India, coal is mainly concentrated in five states – Jharkhand, Odisha, Chhattisgarh, West Bengal, and Madhya Pradesh, which collectively account for 85 per cent of total resources in the country. Jharkhand tops the chart with more than 25 per cent of India's total coal resources (Ministry of Coal, 2019a). India ranks second in global coal production after China, accounting for 9.5 per cent of global production. The production of coal has been steadily increasing in recent years. In the past decade, between 2009–2010 and 2019–2020, production has grown at a compound annual growth rate of 3.2 per cent, with total production increasing from 532 million tonnes in 2009–2010 to 729 million tonnes in 2019–2020 (NSO, 2020a). The Government of India has further set a target of reaching one billion tonnes of production by 2023–2024, which would require the production rate to increase to 8.2 per cent per annum over the next four years (Ministry of Coal, 2019b). In recent years, consumption has increased faster than production and imports have bridged the demand–supply gap (Ministry of Coal, 2020a). Nearly 66 per cent of raw coal is consumed by the power sector, with other key users being cement, iron, and steel (NSO, 2020a).

To accelerate domestic production, the central government has recently taken several steps. In March 2020, India enacted the Mineral Laws Amendment Act 2020, which amends the Coal Mines (Special Provisions) Act of 2015. A key focus of the amendment is to attract private players, including foreign companies, for commercial coal mining (GoI, 2020). In May 2020, the central government further announced an investment of INR 500 billion (US$7 billion) to build coal transportation infrastructure (Ray, 2020) and INR 7 billion (US$97 million) for exploration (Singh, 2020).

Coal-based thermal power accounts for about 55 per cent of total installed capacity but 71 per cent of gross electricity generation (CEA, 2020a). However, coal's dominant role in power generation will weaken gradually due to slower than expected growth in electricity demand and the rapidly falling cost of RE. However, while the share of coal in India's electricity mix is projected to decrease significantly over the next decade, the amount of CO_2 emissions from coal combustion is expected to increase – even if only slightly – due to increasing coal consumption in absolute terms (CEA, 2020b).

The share of coal in overall GDP may not appear to be significant (0.7 per cent) (Spencer et al., 2018), but the taxes and levies from coal constitute an important source of revenue for the centre and several state governments. Between 2013–2014 and 2017–2018, the central government earned INR 2 trillion (US$28 billion) in revenue from Coal India Limited (CIL), the largest state-owned coal mining company producing 83 per cent of India's coal. In the largest coal-producing states, coal assumes an even more important role – 10 per cent and 9 per cent of state revenue in Jharkhand and Chhattisgarh, respectively (Spencer et al., 2018). Another important source of revenue for the government is the Indian Railways. Coal constitutes about 44 per cent of Indian Railways' freight revenue and an even higher share of its profits. This outcome is the result of a deliberately adopted business model that subsidises passenger fare and overcharges freight to fulfil a "social obligation" (Kamboj & Tongia, 2018).

Coal mining is a key source of employment in major coal-producing districts of India. The total formal employment in the state-owned coal mining companies is currently estimated to be around 350,000, which has been decreasing steadily over the years mainly due to mechanisation. Between 2000 and 2014, formal employment declined by about 1.8 per cent per year (CIL, 2018). Outsourcing of coal-mining-related activities enabled by the easing of coal mining laws since the 1990s has led to a mushrooming of private subcontractors and contractual employment. Not only do these contractual jobs pay far less, but they also are not reflected in the official coal company statistics, thus leading to serious underestimation of total coal employment (Chandra, 2020). In addition, a huge population is indirectly dependent on coal as informal workers, scavengers, and the like. Scholars have estimated that the total number of people directly or indirectly dependent on coal for income in the coal belt may be as high as 15 million (Chandra, 2019). A decrease of formal employment in the coal mining sector is expected to continue in the coming years.

Formal employees of the state-owned coal companies benefit from relatively high wages and extensive welfare benefits, including housing, healthcare, education, water supply, etc. (Ministry of Coal, 2020b). In addition to such infrastructure investment, state-owned coal companies have also spent significantly on various social development activities as part of Corporate Social Responsibility (CSR) schemes in the mining regions. For example, CIL spent about INR 30 billion (US$450 million) on CSR projects between 2012–2013 and 2018–2019 (CIL, 2020). While CIL remains a highly profitable company overall, 70 per cent of its mines, especially the underground ones, have become unprofitable (Sehgal & Tongia, 2020). This raises questions about the long-term viability of legacy coal-mining companies in India.

2.6 The Rationale for a Just Transition in the Indian Coal Sector

Discussing a coal phase-out in India is generally perceived as off-limits, with a wide gap between "supporters" and "opponents" of coal. Such polarisation has generally precluded the scope for any collective and deliberative discussion on the future of coal in the face of new realities. Five critical factors compel us to discuss the coal sector in its entirety, including the need for a Just Transition.

2.6.1 Competitiveness of Renewables

The strongest rationale for using coal has been its lower cost and domestic availability. Given coal's availability and the imperative of 100 per cent, country-wide electrification at affordable prices, all negative externalities associated with coal have been overlooked for a long time. However, with a steady decline in the cost of REs, especially solar and wind, brought about by innovation and increased deployment, the cost advantage of coal has been gradually eroding. The installed cost and tariffs of utility-scale solar photovoltaic plants in India have fallen by 85 per cent since 2010, coming down to INR 2.50–3.00/kWh (US$0.035–0.045/kWh) by 2019 (Henbest, 2020). This is already 20–30 per cent cheaper than the cost of power from existing coal plants. This trend is widely expected to continue, with solar power prices expected to fall to around INR 2.00/kWh (US$0.025/kWh) or even lower by 2030, while the cheapest pithead coal power price is likely to rise to INR 4.85/kWh (US$0.07/kWh) by that time (CPI and TERI, 2019).

The biggest "threat" to coal power is from uninterrupted RE and it is now a reality. In a recent auction to supply round-the-clock electricity from RE with battery storage, the average tariff was INR 4.30/kWh (US$0.06/kWh) – the cheapest renewable-plus-battery storage tariff anywhere in the world. At this tariff, renewable-plus-battery storage will out-price most new coal-based power plants (Bhushan, 2020).

Investors are responding to this reality, and it has become increasingly difficult to finance new coal-based power plants in India. Nearly 50 gigawatts (GW) of planned capacity was cancelled in 2019, and many more plants, in the planning stage for years, are not likely to see the light of day (Shearer et al., 2020). The number of reputed international institutional investors formally declaring an end to coal investment in keeping with Paris climate goals is growing by the month. If the government-owned Power Finance Corporation – the most important investor in coal-based power plants in India – stops funding coal power plants, there would hardly be any significant investments in coal power in the country. The Government of India's ambitious intermediate target to create 175 GW of RE capacity by 2022 has given the necessary policy momentum to the RE sector and helped build investor confidence. Since 2016–2017, capacity additions in RE have steadily surpassed that in coal-based thermal power. Even the Central Electricity Authority (CEA) forecasts that the share of RE in total installed power capacity will be nearly twice that of coal by 2029–2030 (CEA, 2020b). The share of gross electricity generation from RE is projected to rise from 21 per cent today to 44 per cent by the end of the decade (CEA, 2020b).

2.6.2 Rising Pollution Control Costs

Both coal-fired power generation and coal mining have several detrimental effects on public health and the environment. To reduce these impacts, environmental standards have been gradually becoming stricter, denting coal power's cost advantage. Coal-based thermal power plants (TPPs) are the largest source of industrial pollution in India, accounting for 60 per cent of suspended particulate matter (SPM) emissions, 50 per cent of sulphur dioxide, 30 per cent of nitrogen oxides, and 80 per cent of mercury emissions (Bhushan et al., 2019b). Inadequate monitoring and poor compliance make matters worse. In January 2020 alone, 34 thermal power stations were issued show-cause notices for non-compliance (Lok Sabha, 2020). Air pollution from coal power plants has major health impacts, not only in the vicinity of the plants but in populations as far as 100 km away. A study conducted by Resources for the Future in 2012 found 650 excess premature deaths per plant in a year at current emissions levels (Bhushan et al., 2015).

Besides air pollution, high water consumption by TPPs and water pollution are also significant issues. India's power plants are estimated to withdraw around 22 billion cubic metres of water, equalling over half of India's non-agricultural water demand (Bhushan et al., 2015). The poor fly ash handling, storage, and disposal systems of TPPs lead to the discharge of ash slurry into natural water bodies causing ecosystem pollution. Nearly 50 per cent of the 170 million tonnes of fly ash generated per year remain unutilised and are dumped into poorly designed and maintained ash ponds. Nearly one billion tonnes of this toxic ash lie in these ponds currently, polluting land, air, and water. Ash dyke breach and the resultant pollution of land and water bodies are quite common in the country (Nandy, 2020). Coal-based industrial activities, such as coal washeries, coke ovens, and soft coke plants, have also been identified by the Central Pollution Control Board as key contributors to the pollution of ambient air, water, and soil in the coal regions (Bhushan & Banerjee, 2015).

Considering the pollution burden, standards on industries and power plants have been progressively made stringent. In 2015, the Ministry of Environment, Forests, and Climate Change (MoEFCC) imposed new standards on coal power plants to limit their emissions of sulphur dioxide, nitrogen oxides, SPM, and mercury (MoEFCC, 2015). While the TPPs have not met the deadline of December 2017, they are regularly given a new deadline to meet these standards (CSE, 2023). The estimated capital expenditure required to meet these standards is about INR 860 billion (US$12 billion). This will add between INR 0.32–0.72/kWh to the existing tariffs (around 9–21 per cent to their average generation tariff), depending on unit size and other factors (Garg et al., 2019). Many older plants will be forced to close because of the tariff increase.

Coal mining has major impacts on forest and forest-based livelihoods. Coal mining accounts for about half of all the forestland diverted for mining. Coal mining also generates a large amount of pollution and waste. Coal mines especially abandoned mines in old coal-mining areas are a significant source of ground and surface water pollution owing to acid-mine drainage. The overburden and waste generated from the mining activities also cause air and water pollution. As a result, most of the top coal-mining areas have been identified as "critically polluted" by the MoEFCC (Bhushan & Banerjee, 2015). A temporary moratorium was imposed on these areas in 2010, and all expansion and new projects were stopped till an action plan was developed to improve environmental quality. While the moratorium has been lifted, these areas now have to implement short-, medium-, and long-term action plans to reduce pollution levels. If they fail to do so, they can be penalised, including the imposition of another moratorium (Aggarwal, 2019).

Overall, as environmental consciousness increases in the country, coal mining and coal-based industries will have to meet higher environmental standards, making them even more uncompetitive to alternative, primarily renewable sources of energy.

2.6.3 Reversing the Resource Curse

Most of India's top coal-mining districts suffer from the classic problem of "resource curse," where the local people have been facing displacement and deprivation due to resource extraction. Coal mining has been controversial over the years due to displacement and resettlement issues. Displacement of people from their lands and livelihoods is one of the worst impacts mining-affected communities have had to endure. While the data on displacement are extremely poor, estimates suggest that over the past 40 years about 87,000 people have been displaced by mining activities of CIL alone. Further, one out of six people displaced belongs to the *Adivasis* or local indigenous community (Reuters, 2016). However, the number is a gross underestimation. In reality, this only includes the people who have lost their lands and were counted as affected by the project, but not the ones that depended on the land for their livelihoods (Bhushan & Zeya Hazra, 2008).

Coal-mining districts are normally economically backward. While coal production from these regions has contributed to the country's growth story, people in these areas have remained impoverished (Bhushan & Zeya Hazra, 2008). A review of India's top 22 mining districts, producing at least 10 million tonnes of coal per year, shows that 50 per cent of the population in most of these districts are "multi-dimensionally poor," exhibiting the very poor status of health, education, and living standards. This is near twice the Indian average of 27.5 per cent (Oxford Poverty and Human Development Initiative,

2020). The average income levels among households are also lower than the Indian average. If rural household income is considered, as most mines are located in rural areas, the highest-earning member's monthly income is less than INR 5,000 (US$69) representative of 85 per cent of rural households in coal-mining districts. The Indian average is 74.5 per cent (Department of Rural Development, 2011). The low-income level suggests that most people do not have a secure income and are involved in low-paid casual labour. The current socio-economic conditions of coal-mining districts are untenable and will have to be reversed to provide jobs and improve the living standards of the population. This will require economic diversification and investment in social sectors.

2.6.4 Urgency of Climate Action

India is one of the most vulnerable countries to the effects of climate change (Eckstein et al., 2020). In 2018, India's economic losses due to climate change were the second-highest globally at about INR 2.7 trillion (US$37.5 billion), roughly 0.36 per cent of the country's GDP (Eckstein et al., 2020). In the same year, a report of the World Bank further noted that climate change will shave off 2.8 per cent of India's per capita GDP by 2050 and worsen the living standards of half of the country's population under the business-as-usual scenario. In the hotspots of Chhattisgarh, Madhya Pradesh, and Maharashtra, the per capita GDP loss could be as high as 9.8 per cent (Mani et al., 2018). Other studies have projected a loss of annual GDP by 2.5–4.5 per cent by 2030 due to heat waves alone (Woetzel et al., 2020). The Department of Economic Affairs (2020) estimates the cumulative costs of climate change adaptation till 2030 at INR 85.6 trillion (US$1.2 trillion), equivalent to nearly 58 per cent of India's GDP in 2019–2020 at constant prices (NSO, 2020b).

Climate change will increase the stress on the country's natural ecosystems, agricultural output, and freshwater resources while also causing escalating damage to infrastructure. These will have severe consequences for the country's food, water, energy security, and public health altogether. In other words, in the absence of rapid, informed, and far-reaching mitigation and adaptation measures, climate change impacts are likely to pose serious challenges to sustaining the country's economic growth and achieving SDG targets (Dhara et al., 2020).

It is in India's interest that GHG emissions are reduced significantly, and effort is made globally to fulfil the Paris Agreement. But the world is nowhere near meeting the target of 1.5°C. In fact, even if countries meet the commitments made under the Paris Agreement, the world is heading for a 3.2°C temperature rise above pre-industrial levels (UNEP, 2019). The Annual Emissions Gap Report of the United Nations, which compares where GHG emissions are headed versus where they need to be, shows that emissions need to fall by 7.6 per cent each year over the next decade if the world is to get back on track towards the goal of limiting temperature increase to close to 1.5°C. This would require all countries to substantially increase their Nationally Determined Contributions (NDCs) bolstered by policies to implement them (UNEP, 2019). Under current policy mechanisms, the country is projected to meet its unconditional NDC target, that is the reduction of emissions intensity of GDP by 33–35 per cent below 2005 levels by 2030 (UNEP, 2019). In fact, India's emissions are likely to be 15 per cent lower than stated in the first set of NDCs (UNEP, 2019). In terms of non-fossil fuels in the electricity mix (installed capacity), the latest estimates by the CEA indicate that the share could be as high as 65 per cent by 2029–2030 (CEA, 2020b). The revised set of NDCs presented by Prime Minister Modi at COP26 in Glasgow is thus only a step towards tapping India's full potential.

It is undeniable that curtailing coal consumption is an absolute necessity for India to reduce its GHG emissions, as it is the single largest source of such emissions in the country. India, therefore, must start planning for its coal phase-out to make a meaningful contribution towards global climate action.

2.6.5 Avoiding Unplanned Mine Closure

While many experts believe that the phase-out of coal power by 2050 is eminently feasible (Bhushan, 2017), a deliberative approach is necessary to avoid socio-economic disruptions in the coal-mining areas, as observed from global experience. Case studies of unplanned mine closure show that closure has been most definitely followed by a surge in unemployment, poverty, and net outmigration from the affected regions. It has also resulted in social protest and political unrest in some parts. For India, the literature on the unplanned closing of mining activities – as limited as it may be – does not tell a very different story (Arondekar & Murthy, 2017; Mishra, 2018).

In fact, coal mines are already closing in the country without any plan for the socio-economic transition of the affected communities. This is especially true for older coal-mining areas in states like Jharkhand, where out of 203 mine leases, 106 are already temporarily or permanently closed (Department of Mines and Geology, 2020). Many of them, especially underground mines, have closed down because they are unprofitable. In fact, about half of Jharkhand's operational mines are also unprofitable and are being cross-subsidised by profitable mines.

The situation is worse at the district level. In Bokaro, 16 of the 25 mines are already closed. Of the nine operational mines, six will exhaust their reserves in the next ten years. In Jamtara, all five mines are closed. A similar situation exists in districts like Hazaribagh and Ramgarh (Department of Mines and Geology, 2020). This exemplifies that the current situation is one of turning a blind eye to an impending crisis. The issue of unplanned closure, which is silently brewing, is highly likely to become a full-blown crisis if we do not address it now. Yet India's coal sector remains largely profitable, even as many districts are burdened with unprofitable mines. This peculiar situation – where profitable and unprofitable mining coexists – necessitates a deliberate approach to Just Transition planning.

2.7 An Indicative Just Transition Framework for India

The Just Transition can be "good economics and good politics," as pointed out by the Global Commission on the Economy and Climate (2018), if the transition away from a coal economy is planned and managed well. This will help mitigate the impacts of climate change and ensure a growth trajectory that is sustainable and equitable while reducing exposure to social and economic disruptions. Evidently, there is no universal, standardised framework for a Just Transition. It is widely accepted that the approach has to be context-specific.

We have reviewed coal transition initiatives and Just Transition programmes of various countries and international organisations, including the Canadian Just Transition Task Force (Government of Canada, 2018), the Guidelines for a Just Transition provided by the International Labour Organization (ILO, 2015), a report by the World Bank Group (2018) on Managing Coal Mine Closure, Strategies for Just Energy Transitions by the International Institute for Sustainable Development (Zinecker et al., 2018), and the Road Map for Just Transition by the Climate Equity Network (Cha et al., 2020). Against this backdrop, a Just Transition in India appears much more complicated, considering the wide diversity of coal-mining districts. For instance, some areas are "mono-industrial" coal towns like Dhanbad or Korba, whereas others have a more diverse economy such as Nagpur. Similarly, the mix of formal and informal coal economy makes a Just Transition in India arguably more complex than in the Western context.

2.7.1 Just Transition Approach

Proposing a framework for Just Transition necessitates understanding what justice means in the context of India's coal-mining areas. As history and local contexts suggest, the mining areas suffer from distributive injustice of resource benefits, economic stratification, and social exclusion. If justice must be ensured in the coal-mining areas, all of these factors need to be accounted for. Considering this, it can be argued that the most suitable approach for Just Transition in the Indian context will be a process of structural reform, which can ensure both distributive and procedural justice for people in the coal-mining areas. Distributive justice will entail planning a transition to ensure fair sharing of costs and benefits among various categories of people dependent on the coal economy and achieve equitable outcomes. Procedural justice mandates an inclusive decision-making process and recognition of collective ownership (just Transition Research Collaborative, 2018). Such an approach to Just Energy Transitions in India's coal sector will also help move away from top–down processes and strengthen bottom–up planning – rooted in the interest of the community – within an inclusive governance system. An indicative Just Transition Framework for India will have two components: (1) pillars and key elements of Just Transition and (2) policies with defined roles of various levels of government. The following two subsections detail these components.

2.7.2 Pillars and Key Elements

A Just Transition framework can be built around the following eight pillars, with their respective key elements as described in Table 2.1.

Table 2.1 Pillars and key elements for Just Transition in India.

Pillars	*Key elements*
1. Strong national- and state-government support	– Develop regional, state, and national plans for coal phase-out. – Establish a coherent policy architecture, where provisions of Just Transition are integrated into national plans and policies, and in policies and plans of the state and local governments. – Establish inter-ministerial and multi-departmental coordination cells for Just Transition. – Facilitate social dialogue throughout the stages of Just Transition (pre-coal mining phase-out, during phase-out, and post-phase-out) to ensure transparent decision-making. – Set up a multi-stakeholder special body/agency for coal mine closure, while entrusting it with the highest stake and responsibility to lead the process. – Provide financial and tax incentives to attract new low-carbon industries (that support local jobs) to secure diverse income and revenue sources.
2. Diverse coalition and local engagement	– Develop a strong and diverse coalition of various actors and stakeholders (including workers, community members, labour unions, environmentalists, businesses, and local government) to work towards a shared vision of Just Transition through coordination and consensus. – Prepare a joint Just Transition proposal and engage with local communities.
3. Effective communication strategy	– Develop a communication strategy to reach out to all stakeholders, from national to local levels, to clearly convey the objective, need, and time frame for Just Transition. – Develop a dedicated portal and information repository.
4. Local support systems	– Open a Just Transition cell at local administrative levels to provide information and engage with formal and informal workers to identify appropriate opportunities to substitute their incomes/livelihoods, as per their skills, adaptive capacity, and scope of mobility.

Pillars	Key elements
5. Economic diversification and social security	– Develop an economic diversification plan, taking into consideration local resources, manpower, and connectivity. – Foster local entrepreneurship through investments in vocational training centres and educational programmes/facilities. – Invest in land and environmental restoration to create long-term economic and social assets.
6. Coal sector transition	– Enable restructuring of CIL and its subsidiaries, for diversification into the solar, wind, energy storage, e-mobility, and the like, that is also aligned to the public purpose of energy security. – Create a special "Workforce Transition Fund" to provide relief measures for displaced workers, such as temporary income support, mobility assistance, an early retirement package, and training and reskilling assistance for younger workers.
7. Social and physical infrastructure development	– Invest in infrastructures, such as education, healthcare, clean water, energy, connectivity, and urban amenities to improve social capital and to attract businesses and investors.
8. Public and private investments	– Identify local, national, and international funding sources and create a dedicated public fund. – Use public funds to attract private investments.

2.7.3 Policies and the Role of Various Levels of Government

Implementing the various pillars of Just Transition will require support from all levels of the government. Based on ground-level engagement in Jharkhand state, key policy directions required to support Just Transition are identified and the respective roles of the central-, state-, and district-level governments are outlined.

2.7.3.1 Central Government

The central government has the most important role in developing supportive policies and programmes for Just Transition. Five major areas in which the role of the Centre will be crucial are the following:

(1) National Just Transition Policy

The Government of India will need to formulate a Just Transition policy as an integral part of India's climate change mitigation strategy and clean energy transition. A Just Transition policy should include the following key elements:

- Coal phase-out policy and plan: defined scale, scope, and timing for capping coal production and regional, state, and national schedules for coal mine closure.

- Coal power plant closure plan: a coal phase-out plan needs to include a complimentary plan for coal-based power plants closure.
- Policies and incentives for fostering clean energy integration: the time frames of phasing out coal-based thermal power generation need to be aligned with the integration of clean energy alternatives.
- Revision of rules and regulations related to the coal industry, including laws related to workers, pensioners, mine closure, and transfer of mine lease area.
- A framework for Just Transition considering the eight pillars specified in Table 2.1 (strong national- and state-government support; diverse coalition and local engagement; effective communication strategy; local support system; economic diversification and social security; coal sector transition; social and physical infrastructure development; public and private investments).

(2) Convergence Plan for National Policies and Programmes

There are important laws and flagship programmes of the central government, which can be augmented, converged, and suitably aligned with the Just Transition policy and its implementation. In fact, proper implementation of these laws and programmes through increased budgetary support and convergence can help to build a solid base for Just Transition immediately. In this regard, the following are some of the important programmes:

(a) *National Mineral Policy and Guidelines*

Proper mine closure and associated land development practices are crucial for improving livelihood and income opportunities in the mining areas. In fact, the restored land can become an economic asset for communities once mining ceases to exist. The National Mineral Policy (Ministry of Mines, 2019) and guidelines of the Ministry of Coal provide directions to this end. The latest guidelines of the Ministry of Coal (2019c) have revised the surety payment for mine closure to INR 900,000/hectare (US$12,000/hectare) for open-cast mines, and INR 150,000/hectare (US$2,000/hectare) for underground mines (Ministry of Coal, 2019c). While this amount might not be sufficient, it certainly provides the seed fund to attract private investments in land reclamation.

(b) *Mahatma Gandhi National Rural Employment Guarantee Act*

Enacted in 2005, the Mahatma Gandhi National Rural Employment Guarantee Act (MGNREGA) is the biggest livelihood support programme for poor and unskilled workers in rural India, assuring wage employment for at least 100 days per household per year. MGNREGA focuses on natural resource management activities such as soil and water conservation, groundwater recharge, land restoration, afforestation, drainage, and irrigation (Ministry of Rural Development, 2018) and can significantly benefit mining-affected regions.

(c) *Forest Rights Act*

The Scheduled Tribes and Other Traditional Forest Dwellers Act of 2006, also referred to as the Forest Rights Act (FRA), emphasises community-based governance of forests. One of the key opportunities for livelihood enhancement through FRA lies in the recognition of community forest resource rights (Ministry of Tribal Affairs, 2014). As most coal-mining districts also have large forest cover, FRA has enormous potential for enhancing livelihood opportunities.

(d) *Aspirational Districts Programme*

The "Transformation of Aspirational Districts" programme launched in 2018 aims to improve the human development indicators of various districts that are considered backward through a concerted, expeditious, and effective effort, focusing on five sectors – health and nutrition, education, agriculture and water resources, financial inclusion and skill development, and infrastructure including roads, access to potable water, electricity, and sanitation facilities. The broad contours of the programme are a convergence of central- and state-government schemes; a collaboration between central-, state-, and district-level officials; and competition among districts to incentivise better performance (NITI Aayog, 2018). Many coal-mining districts are "backward" and therefore already come under the purview of this programme.

(e) *National Rural Livelihood Mission*

The National Rural Livelihood Mission, now known as Deendayal Antyodaya Yojana, is targeted towards poverty alleviation in rural India. The scheme, launched in 2011, aims to increase household income through sustainable livelihood opportunities and improved access to financial services (Ministry of Rural Development, 2021). With its focus on building strong community institutions, the scheme helps poor households to access skilled wage employment opportunities through women's self-help groups and can play a key role in rural economic security in mining-affected regions.

(3) Transitioning of Public Sector Coal Companies

Transitioning of the public sector coal companies will have to happen at two levels. First, the coal business itself will have to be restructured to allow proper and phased closure of mines. Second, these companies will have to transition from being a coal company to an energy company or a mining company. Transitioning of the business will be an important mechanism for diversification of the economy in the coal-mining areas and absorbing a significant proportion of the coal-mining workforce through retraining and reskilling.

CIL, in fact, has been making announcements over the past two years for investing in RE. In 2018, the company said it plans to set up 20 GW of solar projects in the next ten years (Bhaskar, 2018). In August 2020, it was reported that CIL and Neyveli Lignite Corporation Limited (NLC) India Limited will jointly develop solar power assets of 3 GW with a likely investment of around INR 120 billion (US$1.6 billion) (Press Trust of India, 2020). The government has also been deliberating on leveraging CIL's mining expertise to foray into non-fuel minerals mining, including deep-seated and strategic minerals. If India is to build its RE assets and infrastructure, coal company diversification in all of these sectors will be important.

(4) Financial Package

Closing coal mines will have significant socio-economic impacts, including on local businesses and industries, as well as reduced revenue for state governments. These economic losses will have to be substituted in one way or the other. Ideally, a well-planned Just Transition should improve economic opportunities at the state and district levels and enhance revenue sources for the government. But to achieve this, policies to enhance social capital and diversify economic activities must be strengthened, while a financial package for infrastructure and industrial development should be provided. The financial package could include tax incentives, capital, and operational subsidies, infrastructure support, etc.

2.7.3.2 State Government

The state governments will be at the front line for dealing with Just Transition. They will have a vital role to play with regard to the following aspects:

- Building a broad-based consensus and action plan to phase out coal mining and power plants.
- Developing an economic and industrial policy that can help to steer away from coal dependence with respect to industrial development, employment, and government revenue.
- Developing a state-level policy for Just Transition, aligned with the national policy and state development vision.
- Facilitating and implementing the socio-economic transition using state and national policy instruments and financial resources.

A state-level Just Transition policy that is aligned to its vision and action plan can, in fact, be a starting point for facilitating a concerted approach on Just Transition for the coal-mining areas. For example, the Jharkhand state government's "Vision and Action Plan" prioritises economic growth, social upliftment, and livelihood diversification through an output-outcome-oriented approach (Government of Jharkhand, 2016). The three-year action plan, applicable until 2020–2021, had also identified certain outputs against various developmental goals. The vision and action plan of the state is also aligned to the state's Industrial and Investment Promotion Policy (2016) that outlines the blueprint for promoting manufacturing, developing infrastructure, fostering innovation, and creating employment opportunities across sectors for mitigation of poverty and unemployment in the state.

2.7.3.3 District Administration and Local Institutions

A key criterion of Just Transition is local engagement and community dialogue (Cha et al., 2020). An implementation plan needs to be developed at the local level through such engagement. The district administration and local institutions have crucial roles to play in the following two aspects:

(1) Participatory Planning Process

India has constitutional provisions as well as extensive, though not always successful, experience in local planning. The Constitution of India, through the 73rd and 74th Amendments (GoI, 1992), provides a clear mandate for developing local-level plans for economic development and social justice, including an architecture for the development of such plans. Articles 243G and 243W require the development of such plans by Panchayati Raj Institutions (PRIs) – institutions of (self) government for villages – and urban local bodies. Further, Article 243ZD specifies the constitution of a District Planning Committee in every district, who shall be entrusted to consolidate the plans prepared by the panchayats and the municipalities in the district, and accordingly prepare a draft development plan for the district. A detailed guideline for developing district plans is also available (Planning Commission, 2006).

Over the years, a participatory planning process has been encouraged and mandated for programmes that are particularly targeted for alleviating poverty and addressing underdevelopment in rural and distressed areas. Some of the important exercises in this regard have been planned under the Backward Region Grant Fund programme in 2007, the Intensive Participatory Planning Exercise of MGNREGA schemes in 2014, Saansad Adarsh Gram Yojana in 2014, and the Gram Panchayat Development Plans Exercise in 2015 under the 14th Finance Commission grants (Bhushan et al., 2019a). Just Transition planning at the district level can build on these experiences.

(2) Implementation Mechanism and Financial Provision

Implementation of Just Transition policies will require proper administrative mechanisms and fiscal provisions at the district level. A Just Transition task force can be created, based mostly on the existing administrative structure.

The task force should have representatives from various government departments, coal industry, other industries, PRIs, and independent civil society representatives. Concerning funding, the economic package provided by the central and state governments will have to be converged at the district level.

Additionally, the funds of the District Mineral Foundation (DMF) can play a major role in facilitating Just Transition at the district level. DMF was instituted under the Mines and Minerals (Development and Regulation) Amendment Act of 2015 and created trusts in mining districts funded by mineral royalties for the benefit of mining-affected communities. In major coal-mining states and districts, the fund has accrued substantial amounts – nearly INR 100 billion (US$1.4 billion) in Odisha and more than INR 50 billion (US$700 million) each in Jharkhand and Chhattisgarh (Ministry of Mines, 2020). Though fund utilisation remains dismal at present (Banerjee et al., 2018; Banerjee, 2020), proper planning and administration can channel the fund's enormous potential to support livelihood diversification and improve social infrastructure. In fact, the Pradhan Mantri Khanij Khestra Kalyan Yojana guidelines, which have been aligned to DMFs, specify the need for creating an endowment fund from the DMF budget to support sustainable livelihoods in mining areas post-mine closure. The Pradhan Mantri Khanij Khestra Kalyan Yojana guidelines can be further amended to include Just Transition as an important component of DMF investments.

2.8 Conclusion

The above analysis has attempted to demonstrate why adopting a proactive and comprehensive Just Transition planning framework for the coal sector, supported by appropriate policy measures is in India's interests. While "coal phase-out" is still considered a "taboo" in government policy circles, the reality is that coal mining closures are already happening and will accelerate in the years ahead because of inescapable economic and environmental constraints. Policymakers need to realise that (1) unplanned collapse of such an important sector will have devastating socio-economic and ecological consequences and (2) Just Transition planning and execution take decades as the international experience has shown. After a point, it will be too late to avoid a disorderly collapse of the sector. Therefore, a multi-stakeholder-based process must commence in earnest without delay. The good news is that many elements of a future Just Transition framework already exist in the form of policies and programmes at various levels of government. The convergence and streamlining under a Just Transition "umbrella" can go a long way towards building momentum in a relatively short time span.

The world needs bold, transformative solutions that can bend the present trajectories of fossil fuel consumption, including coal, for addressing the escalating threats of climate change. To ensure that developing countries reduce their coal consumption, a "Global Support Programme" for Just Transition is needed to fund a fair transition, including energy transition investments. The Silesia Declaration on "Solidarity and Just Transition" points out the need for "encouraging support for developing countries to promote low greenhouse gas emission and sustainable economic activities that will create quality jobs in regions, cities and non-urban areas" (UN, 2018). While India was not a party to this declaration at present, it has a major stake in defining the global debate on Just Transition. India should, therefore, take leadership in facilitating the development of a "Global Support Programme" for Just Transition, in parallel with its domestic initiatives.

Note

1 Signatories from the Global South include Argentina, Eswatini, Fiji, Gabon, Honduras, Indonesia, Madagascar, Mexico, Mongolia, Namibia, Nauru, Nepal, Nigeria, Palestine, Peru, Thailand, Togo, and Uruguay.

References

Aggarwal, M. (2019). Develop time-bound action plans for critically polluted areas: MoEFCC tells state governments. *Mongabay*. 12.11.2019. Retrieved from: https://india.mongabay.com/2019/11/develop-time-bound-action-plans-for-critically-polluted-areas-moefcc-tells-state-governments/ (Last accessed: 24.09.2021)

Arondekar, S.D. & Murthy, I.B. (2017). A study on the socio-economic impact of mining ban on the households in Goa's mining belt. *SDMIMD Journal of Management*, 8(1): 11–19. https://doi.org/10.18311/sdmimd/2017/15715

Banerjee, S. (2020). *District Mineral Foundation Funds Crucial Resource for Ensuring Income Security in Mining Areas Post COVID-19*. Brookings Institution India Centre. Retrieved from: www.brookings.edu/blog/up-front/2020/05/06/district-mineral-foundation-funds-crucial-resource-for-ensuring-income-security-in-mining-areas-post-covid-19/ (Last accessed: 30.08.2021)

Banerjee, S., Shalya, C., Valaparla, S., Ranjan, R., Dhingra, A. & Sarangi, A. (2018). *People First: District Mineral Foundation (DMF) Status Report 2018*. Centre for Science and Environment, New Delhi.

Bhaskar, U. (2018). Coal India to set up 20,000 MW of solar power projects in 10 years. *Livemint*. 30.03.2018. Retrieved from: www.livemint.com/Industry/vNIicVhgfv7UzOrKkj67dP/Coal-India-to-set-up-20000MW-of-solar-power-projects-in-nex.html (Last accessed: 01.08.2021)

Bhushan, C. (2017). The last coal power plant in India can be closed by 2050. *Down To Earth*. 07.11.2017. Retrieved from: www.downtoearth.org.in/coverage/energy/the-end-of-coal-58909 (Last accessed: 15.08.2021)

Bhushan, C. (2020). Here why energy sector embrace technological disruptions and use them as an opportunity. *Financial Express*. 14.02.2020. Retrieved from: www.financialexpress.com/opinion/here-why-energy-sector-embrace-technological-disruptions-and-use-them-as-an-opportunity/1867585/ (Last accessed: 18.08.2021)

Bhushan, C. & Banerjee, S. (2015). *Losing Solid Ground: MMDR Amendment Act, 2015 and the State of the Mining Sector in India*. Centre for Science and Environment, New Delhi.

Bhushan, C., Banerjee, S. & Shalya, C. (2019a). *An Inclusive and Outcome-Oriented Framework for DMF Planning*. Centre for Science and Environment, New Delhi. Retrieved from: www.cseindia.org/an-inclusive-and-outcome-oriented-framework-for-dmf-planning-9255 (Last accessed: 01.09.2021)

Bhushan, C., Bhati, P., Sanjeev, K., Sangeetha, A., Siddhartha, S., Ramanathan, S. & Rudra, A. (2015). *Heat on Power: Green Rating of Coal-based Power Plants*. Centre for Science and Environment, New Delhi.

Bhushan, C., Bhati, P., Sreenivasan, P., Singh, M., Jhawar, P., Miriam Koshy, S. & Sambyal, S. (2019b). *The State of Renewable Energy in India*. Centre for Science and Environment, New Delhi.

Bhushan, C. & Zeya Hazra, M. (2008). *Rich Land Poor People: Is Sustainable Mining Possible?* Centre for Science and Environment, New Delhi.

BP (2020). *Statistical Review of World Energy*, 69th edition. British Petroleum. Retrieved from: www.bp.com/content/dam/bp/business-sites/en/global/corporate/pdfs/energy-economics/statistical-review/bp-stats-review-2020-full-report.pdf (Last accessed: 29.07.2021)

CEA (2020a). *All India Electricity Statistics – General Review 2020*. Central Electricity Authority. Retrieved from: www.indiaenvironmentportal.org.in/files/file/all%20india%20electricity%20statistics%202020.pdf (Last accessed: 04.08.2021)

CEA (2020b). *Draft Report on Optimal Generation Capacity Mix for 2029–30*. Central Electricity Authority, Ministry of Power, Government of India. Retrieved from: http://cea.nic.in/reports/others/planning/irp/Optimal_generation_mix_report.pdf (Last accessed: 08.08.2021)

Cha, M., Pastor, M., Wander, M., Sadd, J. & Morello-Frosch, R. (2020). *A Roadmap to an Equitable Low-Carbon Future: Four Pillars for a Just Transition*. The Climate Equity Network. Retrieved from: https://dornsife.usc.edu/assets/sites/242/docs/JUST_TRANSITION_Report_FINAL_12-19.pdf (Last accessed: 13.07.2021)

Chandra, R. (2019). A coal commission for India. *The Indian Express*. 03.01.2019. Retrieved from: https://indianexpress.com/article/opinion/columns/a-coal-commission-for-india-5520686/ (Last accessed: 05.08.2021)

Chandra, R. (2020). India's coal industry and CIL in historical perspective. In: Tongia, R., Sehgal, A. & Kamboj, P. (Eds.), *Future of Coal in India: Smooth Transition or Bumpy Road Ahead?* Notion Press, Chennai, India.

CIL (2018). *Coal Vision Document 2030. Stakeholders' Consultation*. Coal India Limited. Retrieved from: www.indiaenvironmentportal.org.in/files/file/Coal_Vision_2030.pdf (Last accessed: 10.08.2021)

CIL (2020). *Annual Reports*. Coal India Limited. Government of India Undertaking. Retrieved from: https://coal.gov.in/en/public-information/reports/annual-reports (Last accessed: 18.01.2024)

Climate Justice Alliance (2018). *Just Transition Principles*. Retrieved from: https://climatejusticealliance.org/wp-content/uploads/2018/06/CJA_JustTransition_Principles_final_hi-rez.pdf (Last accessed: 23.07.2021)

CPI and TERI (2019). *Accelerating India's Transition to Renewables. Energy Transitions Commission*. Climate Policy Initiative and The Energy and Resources Institute. Retrieved from: www.teriin.org/sites/default/files/files/etc-key-messages-summary.pdf (Last accessed: 02.08.2021)

CSE (2023). *Coal-Based Thermal Power Plants Irgnoring Pollution Control Measures, Responsible for Aggravating Air Pollution in Delhi-NCR*. Centre for Science and Environment, New Delhi. Retrieved from: https://www.cseindia.org/coal-based-thermal-power-plants-ignoring-pollution-control-measures-responsible-11969 (Last accessed: 25.01.2024)

Department of Economic Affairs (2020). *Report of the Sub-Committee for the Assessment of the Financial Requirements for Implementing India's Nationally Determined Contributions (NDC)*. Ministry of Finance, Government of India. Retrieved from: https://dea.gov.in/sites/default/files/Sub%20Committee%20Report%20Final.pdf (Last accessed: 18.08.2021)

Department of Mines and Geology (2020). *Government of Jharkhand*. Retrieved from: http://jharkhandminerals.gov.in/ (Last accessed: 16.08.2021)

Department of Rural Development (2011). *Socio Economic Caste Census 2011*. Ministry of Rural Development, Government of India, New Delhi.

Dhara, C., Krishnan, R. & Niyogi, D. (2020). Possible climate change impacts and policy-relevant messages. In: Krishnan, R., Sanjay, J., Gnanaseelan, C., Mujumdar,

M., Kulkarni, A. & Chakraborty, S. (Eds.), *Assessment of Climate Change over the Indian Region*. Springer, Singapore.

EC (2020a). *Coal Regions in Transition*. European Commission. Retrieved from: https://ec.europa.eu/energy/topics/oil-gas-and-coal/EU-coal-regions/coal-regions-transition_en (Last accessed: 20.07.2021)

EC (2020b). *The Just Transition Mechanism: Making Sure No One is Left Behind*. European Commission. Retrieved from: https://ec.europa.eu/commission/presscorner/detail/en/FS_20_39 (Last accessed: 20.07.2021)

Eckstein, D., Künzel, V., Schäfer, L. & Winges, M. (2020). *Global Climate Risk Index 2020*. Germanwatch. Retrieved from: https://germanwatch.org/sites/germanwatch.org/files/20-2-01e%20Global%20Climate%20Risk%20Index%202020_10.pdf (Last accessed: 22.07.2021)

Garg, V., Narayanaswamy, D., Ganesan, K. & Viswanathan, B. (2019). *India's Energy Transition: The Cost of Meeting Air Pollution Standards in the Coal-fired Electricity Sector*. International Institute for Sustainable Development. Retrieved from: www.iisd.org/system/files/2020-08/india-energy-transition-air-pollution-standards.pdf (Last accessed: 03.08.2021)

The German Coal Commission (2019). *A Roadmap for a Just Transition from Coal to Renewables*. Retrieved from: www.agora-energiewende.de/en/publications/the-german-coal-commission/ (Last accessed: 21.07.2021)

Global Commission on the Economy and Climate (2018). *Unlocking the Inclusive Growth Story of the 21st Century: Accelerating Climate Action in Urgent Times*. New Climate Economy. Retrieved from: https://newclimateeconomy.report/2018/ (Last accessed: 22.07.2021)

GoI (1992). *The Constitution (Seventy-third and Seventy-fourth Amendment) Act, 1992*. Retrieved from: https://www.india.gov.in/my-government/constitution-india/amendments/constitution-india-seventy-third-amendment-act-1992; https://www.india.gov.in/my-government/constitution-india/amendments/constitution-india-seventy-fourth-amendment-act-1992 (Last accessed: 18.01.2024)

GoI (2020). *The Mineral Laws Amendment Act*. Government of India. Retrieved from: https://coal.nic.in/sites/upload_files/coal/files/curentnotices/gazette%20biil1703 2020.pdf (Last accessed: 04.08.2021)

Government of Canada (2018). *A Just and Fair Transition for Canadian Coal Power Workers and Communities. Report by the Task Force on Just Transition for Canadian Coal Power Workers and Communities*. Retrieved from: https://publications.gc.ca/collections/collection_2019/eccc/En4-361-2019-eng.pdf (Last accessed: 23.07.2021)

Government of Jharkhand (2016). *Jharkhand Vision and Action Plan 2021*. Retrieved from: https://finance-jharkhand.gov.in/pdf/Executive_Summary_Jharkhand_Vision_and_Action_Pln.pdf (Last accessed: 29.08.2021)

Henbest, S. (2020). The first phase of the transition is about electricity, not primary energy. *Bloomberg NEF*. 28.01.2020. Retrieved from: https://about.bnef.com/blog/the-first-phase-of-the-transition-is-about-electricity-not-primary-energy/ (Last accessed: 06.08.2021)

IEA (2016). *World Energy Outlook 2016*. International Energy Agency, Paris. Retrieved from: www.iea.org/reports/world-energy-outlook-2016 (Last accessed: 18.07.2021)

IEA (2018). *World Energy Outlook 2018*. International Energy Agency, Paris. Retrieved from: www.iea.org/reports/world-energy-outlook-2018 (Last accessed: 18.07.2021)

IEA (2019). *World Energy Investment 2019*. International Energy Agency, Paris. Retrieved from: www.iea.org/reports/world-energy-investment-2019/power-sector (Last accessed: 20.07.2021)

ILO (2015). *Guidelines for a Just Transition towards Environmentally Sustainable Economies and Societies for All*. International Labor Organization. Retrieved from: www.ilo.org/wcmsp5/groups/public/-ed_emp/-emp_ent/documents/publication/wcms_432859.pdf (Last accessed: 20.07.2021)

IPCC (2018). *Global Warming of 1.5°C: An IPCC Special Report on the Impacts of Global Warming of 1.5°C above Pre-Industrial Levels*. Intergovernmental Panel on Climate Change. Retrieved from: www.ipcc.ch/site/assets/uploads/sites/2/2019/06/SR15_Full_Report_Low_Res.pdf (Last accessed: 15.07.2021)

ITUC (2017). *Just Transition – Where Are We Now and What's Next? A Guide to National Policies and International Climate Governance*. International Trade Union Confederation. Retrieved from: www.ituc-csi.org/IMG/pdf/ituc_climate_justice_frontline_briefing_2017.pdf (Last accessed: 25.07.2021)

Just Transition Alliance (n.d.). *What is Just Transition*? Retrieved from: http://jtalliance.org/what-is-just-transition/ (Last accessed: 25.07.2021)

Just Transition Centre (2019). *Just Transition in Action: Union Experiences and Lessons from Canada, Germany, New Zealand, Norway, Nigeria and Spain*. Retrieved from: www.ituc-csi.org/IMG/pdf/191120_-_just_transition_case_studies.pdf (Last accessed: 24.07.2021)

Just Transition Research Collaborative (2018). *Mapping Just Transition(s) to a Low-Carbon World*. United Nations Research Institute for Social Development (UNRISD). Retrieved from: www.unrisd.org/80256B3C005BCCF9/(httpPublications)/9B3F4F10301092C7C12583530035C2A5?OpenDocument (Last accessed: 26.07.2021)

Kamboj, P. & Tongia, R. (2018). *Indian Railways and Coal: An Unsustainable Interdependency*. Brookings Institution India Centre, New Delhi. Retrieved from: www.brookings.edu/wp-content/uploads/2018/07/Railways-and-coal.pdf (Last accessed: 09.08.2021)

Lok Sabha (2020). *Unstarred Question No. 3401*. Government of India. 13.03.2020. Retrieved from: https://legalaffairs.gov.in/sites/default/files/AU3401.pdf (Last accessed 18.01.2024)

Mani, M., Bandyopadhyay, S., Chonabayashi, S., Markandya, A. & Mosier, T. (2018). *South Asia's Hotspots: The Impact of Temperature and Precipitation Changes on Living Standards*. The World Bank Group. Retrieved from: https://openknowledge.worldbank.org/bitstream/handle/10986/28723/9781464811555.pdf?sequence=5&isAllowed=y (Last accessed: 30.08.2021)

Ministry of Coal (2019a). *State-Wise Geological Resources*. Government of India. Retrieved from: https://coal.nic.in/content/coal-reserves (Last accessed: 13.09.2021)

Ministry of Coal (2019b). *Coal India to Produce One Billion Tonnes of Coal by 2024*. Government of India. Retrieved from: https://pib.gov.in/Pressreleaseshare.aspx?PRID=1590009 (Last accessed: 14.09.2021)

Ministry of Coal (2019c). *Guidelines for Preparation of Mining Plan for the Coal and Lignite Blocks*. Government of India. Retrieved from: https://coal.gov.in/sites/default/files/2020-01/Guidelines-for-Mining-plan-Coal_16122019.pdf (Last accessed: 17.09.2021)

Ministry of Coal (2020a). *Coal Production and Supplies*. Government of India. Retrieved from: https://coal.nic.in/content/production-and-supplies (Last accessed: 22.09.2021)

Ministry of Coal (2020b). *Annual Report 2019–2020*. Government of India. Retrieved from: https://coal.nic.in/sites/upload_files/coal/files/coalupload/AnnualReport2019-20/Chapter5-en.pdf (Last accessed: 24.09.2021)

Ministry of Mines (2019). *National Mineral Policy, 2019*. Government of India. Retrieved from: https://mines.gov.in/writereaddata/Content/NMP12032019.pdf (Last accessed: 26.09.2021)

Ministry of Mines (2020). *DMF Fund Status up to May 2020*. Government of India. Retrieved from: www.mines.gov.in/writereaddata/Content/DMF%20Fund%20Status_01072020.pdf (Last accessed: 01.10.2021)

Ministry of Rural Development (2018). *Mahatma Gandhi National Rural Employment Guarantee Act: Report on NRM Related Works*. Government of India. Retrieved from: https://mnregaweb4.nic.in/netnrega/nrm_works_report.aspx?lflag=eng&fin_year=2017-2018&source=national&labels=labels&Digest=cT/J7ChEq5LOfEr-0AmsuAQ (Last accessed: 05.10.2021)

Ministry of Rural Development (2021). *Deendayal Antyogaya Yojana-NRLM*. Government of India. Retrieved from: https://aajeevika.gov.in/en/content/welcome-deendayal-antyodaya-yojana-nrlm (Last accessed: 05.10.2021)

Ministry of Statistics and Programme Implementation (2022). *Energy Statistics India – 2022*. National Statistical Office, Government of India. Retrieved from: https://mospi.gov.in/documents/213904/1606151//Energy%20Statistics%20India%2020221644825594802.pdf/aed59aac-4d5a-995b-1232-bb68397cd873 (Last accessed: 11.10.2022)

Ministry of Tribal Affairs (2014). *The Scheduled Tribes and Other Traditional Forest Dwellers (Recognition of Forest Rights) Act, 2006*. Government of India. Retrieved from: https://tribal.nic.in/FRA/data/FRARulesBook.pdf (Last accessed: 05.10.2021)

Mishra, S. (2018). Mine closures and the issue of livelihood. *Economic & Political Weekly*, 53(42).

MoEFCC (2015). *India's Intended Nationally Determined Contribution: Working towards Climate Justice*. Submissions of the Ministry of Environment, Forest, and Climate Change (MoEFCC) to the UN Framework Convention on Climate Change (UNFCCC). Retrieved from: https://www4.unfccc.int/sites/ndcstaging/PublishedDocuments/India%20First/INDIA%20INDC%20TO%20UNFCCC.pdf (Last accessed: 11.08.2021)

Morena, E., Krause, D. & Stevis, D. (Eds.) (2019). *Just Transitions: Social Justice in the Shift Towards a Low-Carbon World*. Pluto Press, London.

Nandy, J. (2020). Third ash-pond disaster in Singrauli in one year. *Hindustan Times*. 11.04.2020. Retrieved from: www.hindustantimes.com/india-news/third-ash-pond-disaster-in-singrauli-in-one-year/story-w9mksJAao8DcsTrkJijUbN.html (Last accessed: 12.08.2021)

NITI Aayog (2018). *Aspirational Districts: Unlocking Potential*. Government of India. Retrieved from: https://niti.gov.in/sites/default/files/2018-12/Aspirational Districts-Book.pdf (Last accessed: 04.09.2021)

NSO (2020a). *Energy Statistics 2020*. National Statistical Office, Ministry of Statistics and Programme Implementation, Government of India. Retrieved from: www.mospi.gov.in/sites/default/files/publication_reports/ES_2020_240420m.pdf (Last accessed: 11.08.2021)

NSO (2020b). *Provisional Estimates of Annual National Income 2019–2020 and Quarterly Estimates of Gross Domestic Product for the Fourth Quarter of 2019–20*. Press Note, National Statistical Office, Ministry of Statistics and Programme Implementation, Government of India. Retrieved from: www.mospi.gov.in/sites/default/files/press_release/PRESS%20NOTE%20PE%20and%20Q4%20estimates%20of%20GDP.pdf (Last accessed: 29.07.2021)

Oxford Poverty and Human Development Initiative (2020). *Country Level Analysis: India*. University of Oxford. Retrieved from: https://ophi.org.uk/multidimensional-poverty-index/country-level/ (Last accessed: 01.08.2021)

Pai, S., Harrison, K. & Zerrifi, H. (2020). *A Systematic Review of the Key Elements of a Just Transition for Fossil Fuel Workers. Clean Economy Working Paper Series: WP 20-04*. Smart Prosperity Institute. Retrieved from: https://institute.smartprosperity.ca/sites/default/files/transitionforfossilfuelworkers.pdf (Last accessed: 20.07.2021)

Planning Commission (2006). *Guidelines for District Plans in the Eleventh Five Year Plan*. Government of India. Retrieved from: https://niti.gov.in/planningcommission.gov.in/docs/plans/stateplan/sp_guideDP.pdf (Last accessed: 01.08.2021)

PPCA (2020). *PPCA Members*. Powering Past Coal Alliance. Retrieved from: https://poweringpastcoal.org/about/members (Last accessed: 03.08.2021)

Press Trust of India (2020). CIL, NLC India to invest Rs 12,000 cr to jointly develop solar power assets. *Business Standard*. 04.07.2020. Retrieved from: www.business-standard.com/article/companies/cil-nlc-india-to-invest-rs-12-000-cr-to-jointly-develop-solar-power-assets-120070400828_1.html (Last accessed: 16.08.2021)

Ray, M. (2020). Govt to spend Rs 50,000 crore to develop coal sector: Nirmala Sitharaman. *Hindustan Times*. 16.05.2020. Retrieved from: www.hindustantimes.com/india-news/govt-to-spend-rs-50-000-crore-to-develop-coal-sector-nirmala-sitharaman/story-Aivg5VKqY5oYDeIVmtbq0J.html (Last accessed: 18.08.2021)

Reuters (2016). Rampant coal mining: One in six of 87,000 Indians displaced is Adivasi, Amnesty says. *Firstpost*. 13.07.2016. Retrieved from: www.firstpost.com/business/rampant-coal-mining-one-in-six-of-87000-indians-displaced-is-adivasi-amnesty-says-2891426.html (Last accessed: 29.07.2021)

Sehgal, A. & Tongia, R. (2020). Coal pricing and allocations in India. In: Tongia, R., Sehgal, A. & Kamboj, P. (Eds.), *Future of Coal in India: Smooth Transition or Bumpy Road Ahead?* Notion Press, Chennai, India.

Shearer, C., Myllyvirta, L., Yu, A., Aitken, G, Mathew-Shah, N., Dallos, G. & Nace, T. (2020). *Boom and Bust 2020: Tracking the Global Coal Plant Pipeline*. Report by Global Energy Monitor, Sierra Club, Greenpeace International and Centre for Research on Energy and Clean Air.

Singh, S. (2020). Budget 2020: Coal India's capital outlay seen rising 18 per cent to Rs 9,500 crore. *The Economic Times*. 01.02.2020. Retrieved from: https://energy.economictimes.indiatimes.com/news/coal/budget-2020-coal-indias-capital-outlay-seen-rising-18-per-cent-to-rs-9500-crore/73841128 (Last accessed: 22.08.2021)

Spencer, T., Pachouri, R., Renjith, G. & Vohra, S. (2018). *Coal Transition in India*. The Energy and Resources Institute. Retrieved from: www.teriin.org/sites/default/files/2018-12/Coal-Transition-in-India.pdf (Last accessed: 30.08.2021)

Stevis, D., Uzzell, D. & Räthzel, N. (2018). The labour-nature relationship: Varieties of labour environmentalism. *Globalizations*, 15(4): 439–453.

TS (2021). Der Koalitionsvertrag 2021. *Der Tagesspiegel*. 29.11.2021. Retrieved from: www.tagesspiegel.de/politik/ampel-steht-der-koalitionsvertrag-2021-als-pdf-zum-download/27829980.html (Last accessed: 07.12.2021)

UN (2018). *Solidarity and Just Transition Silesia Declaration*. United Nations Climate Change Conference: COP 24, Katowice, Poland. Retrieved from: https://cop24.gov.pl/fileadmin/user_upload/Solidarity_and_Just_Transition_Silesia_Declaration_2_.pdf (Last accessed: 06.08.2021)

UN (2020). *Net-Zero Emissions must be met by 2050 or Covid-19 Impact on Global Economies will Pale Beside Climate Crisis, Secretary-General tells Finance Summit*. United Nations Press Release. 12.11.2020. Retrieved from: www.un.org/press/en/2020/sgsm20411.doc.htm (Last accessed: 30.11.2021)

UNEP (2019). *Emissions Gap Report 2019*. United Nations Environment Programme. Retrieved from: www.unenvironment.org/resources/emissions-gap-report-2019 (Last accessed: 23.08.2021)

UNFCCC (2015). *Paris Agreement*. United Nations Convention on Climate Change. Retrieved from: https://unfccc.int/files/essential_background/convention/application/pdf/english_paris_agreement.pdf (Last accessed: 15.07.2021)

UNRISD (2018). *Mapping Just Transition(s) to a Low Carbon World. Just Transition Research Collaborative*. United Nations Research Institute for Social Development. Retrieved from: www.unrisd.org/80256B3C005BCCF9/(httpPublications)/9B3F4F10301092C7C12583530035C2A5?OpenDocument (Last accessed: 28.07.2021)

USEIA (2018). *Annual Energy Outlook 2018 with Projections to 2050*. US Energy Information Administration, U.S. Department of Energy, Washington, DC.

Retrieved from: www.eia.gov/outlooks/archive/aeo18/pdf/AEO2018.pdf (Last accessed: 10.08.2021)

Woetzel, J., Pinner, D., Samandari, H., Gupta, R., Engel, H., Krishnan, M. & Carter, P. (2020). *Will India Get Too Hot to Work?* McKinsey Global Institute. Retrieved from: www.mckinsey.com/business-functions/sustainability/our-insights/will-india-get-too-hot-to-work (Last accessed: 05.08.2021)

World Bank (2018). *Managing Coal Mine Closure: Achieving a Just Transition for All.* The World Bank Group. Retrieved from: www.worldbank.org/en/topic/extractiveindustries/publication/managing-coal-mine-closure (Last accessed: 27.07.2021)

Zinecker, A., Gass, P., Gerasimchuk, I., Jain, P., Moerenhout, T., Oharenko, Y., Suharsono, R.A. & Beaton, C. (2018). *Real People, Real Change – Strategies for Just Energy Transitions.* International Institute for Sustainable Development. Retrieved from: www.iisd.org/system/files/publications/real-people-change-strategies-just-energy-transitions.pdf (Last accessed: 28.07.2021)

3 Gas-Based Economy

A Road Towards a Prosperous and Cleaner India

Akhil Mehrotra, Aayushi Agarwal, and Annika Bose Styczynski

3.1 Introduction

There is a high positive correlation between the consumption of energy and the economic development of any country (Stjepanović, 2018). The energy consumption in developing countries has nearly doubled in the past 15 years and is estimated to increase by another 30 per cent in the next 15 years (Benoit, 2019). This projected level of growth in energy consumption has triggered a series of discussions and raised serious concerns about climate change globally, Just Transitions, and sustainable development more broadly. Historically, the fast-paced rise in global energy consumption has derived from the increased use of fossil fuels (Smil, 2017; BP Statistical Review; Ritchie, 2019). The consumption of fossil fuels is the primary source for unseen levels of greenhouse gas (GHG) emissions in the atmosphere leading to harmful ecological imbalances. Drawing from the experience of advanced Organization for Economic Cooperation and Development economies, there is an emerging view that the growth of an economy and addressing climate concerns are not necessarily discordant (IEA, 2021; Howarth et al., 2017). Such decarbonised growth, however, would require a greater focus on energy productivity and the use of cleaner fuels.

This being said, the challenges for India are dual: a faster gross domestic product (GDP) growth rate along with a focus on a decarbonised energy mix. Today, India's primary energy consumption is dominated by fossil fuels like coal (57 per cent) and oil (27 per cent) (BP, 2023). They are comparatively inexpensive, mostly locally produced or easily accessible through evolved transmission and distribution channels. The demand for energy in India is expected to grow at 4–5 per cent in the medium term even with a modest 3–4 per cent growth in GDP. At the same time, the targets of India's Nationally Determined Contributions (NDCs) to the United Nations Framework Convention on Climate Change (UNFCCC) such as reducing GHG emission intensity by 45 per cent by 2030 have to be met if not outperformed. Hence, the additional energy required by India needs to be not just affordable and accessible but significantly cleaner.

DOI: 10.4324/9781003281818-3

The central government has repeatedly announced to increase the share of natural gas from the current 6 per cent to 15 per cent in the primary energy mix of the country by 2030. Prime Minister Sh. Narendra Modi remarked at the Fourth Annual Energy Forum in October 2020 that India's energy plan will have seven key drivers, the first one being to accelerate efforts to move towards a gas-based economy (PIB, 2020). This was re-emphasised at the India Energy Week 2023 by Minister of Petroleum and Natural Gas, Sh. Hardeep Singh Puri, who stated that India "will remain focused on energy efficiencies, laying impetus on fuels of the future including biofuels and hydrogen, and increasing the use of renewable energies etc. At the same time, we are taking transformative steps to increase domestic exploration and production of traditional hydrocarbons," the cleanest of which being natural gas (Gopi, 2023).

Natural gas has many applications in industry, buildings, power generation, or mobility purposes. However, over the past decade, both industry and government have struggled to bring the share of natural gas to double-digit levels. Multiple structural and systemic issues of the energy sector as a whole and the gas sector specifically account for this backlog. Addressing these issues through mechanisms detailed in later parts of the chapter would not only ensure that gas as an energy source becomes a viable constituent of the Indian energy mix but also help in achieving targets on emission reduction and be conducive to an era of sustainable development.

3.2 India in the Global Economy

Global GDP was reported at US$87.5 trillion in 2019 (O'Neill, 2021). Pre-COVID-19, India reached the status of the fastest-growing trillion-dollar economy and ranked fifth in terms of nominal GDP at US$2.94 trillion. Back in 1980, the Indian economy stood at just US$189.43 billion and ranked 13th in global comparison. Over the past four decades, its size has not only multiplied by a factor of more than 15, but it also earned its position among the top five global economies (World Bank, n.d.). This is a remarkable achievement even if India's population figures drive down nominal GDP to a mere US$2,170 per capita. Over the next decade, a high growth trajectory has been projected for India with ambitious targets of becoming a US$7 trillion to US$10 trillion economy and ranking third in the world by 2030 (Sharma, 2020; Gupta, 2019; PWC, 2014). Historically, economic expansion strongly correlates with the use of fossil fuels. Figure 3.1 illustrates the predominance of conventional fuels in global primary energy consumption from 1800 to 2022.

When we look at this narrative from an Indian perspective, traditional economic development is equally grounded in fossil fuel consumption. In the market liberalisation phase starting in 1991 till the end of Manmohan Singh's prime ministership in 2014, India's GDP grew by more than 650 per cent,

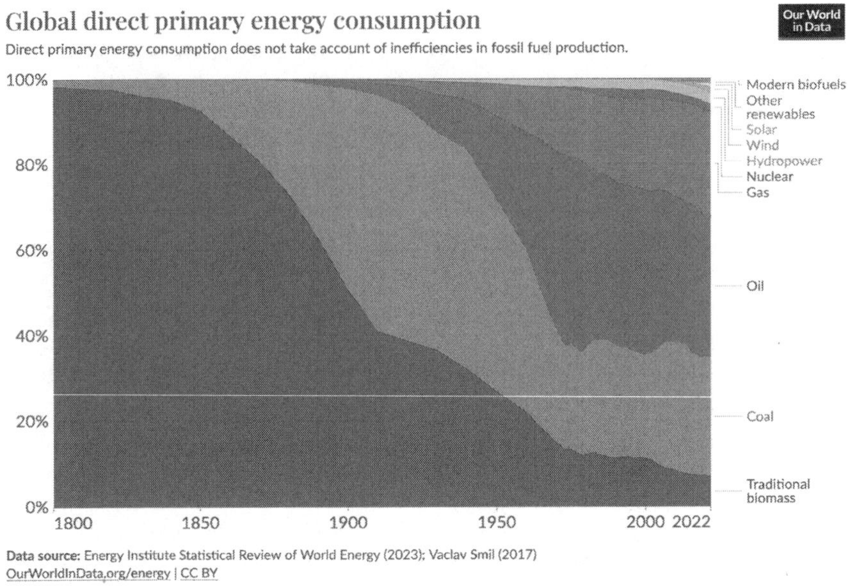

Figure 3.1 Global direct primary energy consumption since 1800.

Source: Our World in Data (2023)

while GHG emissions increased by 182 per cent (World Bank, n.d.; Climate Watch, 2021). With that, India had a much higher increase in emissions relative to GDP growth than China, the United States, Europe, or the world average. By contrast, similar to emission profiles of other countries, the energy sector is the single largest contributor of GHG emissions in India. According to USAID (2018), the Indian energy industry accounted for 68.7 per cent of total GHG emissions. Within the energy sector, electricity and heat generation (49 per cent), manufacturing and construction (24 per cent) followed by other fuel combustion (14 per cent) and transportation (10 per cent) account for GHG emission generation.

3.2.1 Energy Profile of India

As per 2021 data, the dependency on coal in primary energy consumption amounted to 57 per cent (see Figure 3.2) and even 74 per cent in electricity generation (see Figure 3.3), which explains why India today is the third-largest emitter of GHGs after China and the United States (BP, 2021; Timperley, 2019) The high dependency on coal can be attributed to the secured and steady supply of the resource in combination with the lower cost of energy production.

Both Figures 3.2 and 3.3 also show that natural gas plays a minor role in India's energy system as of now and when compared with global figures.

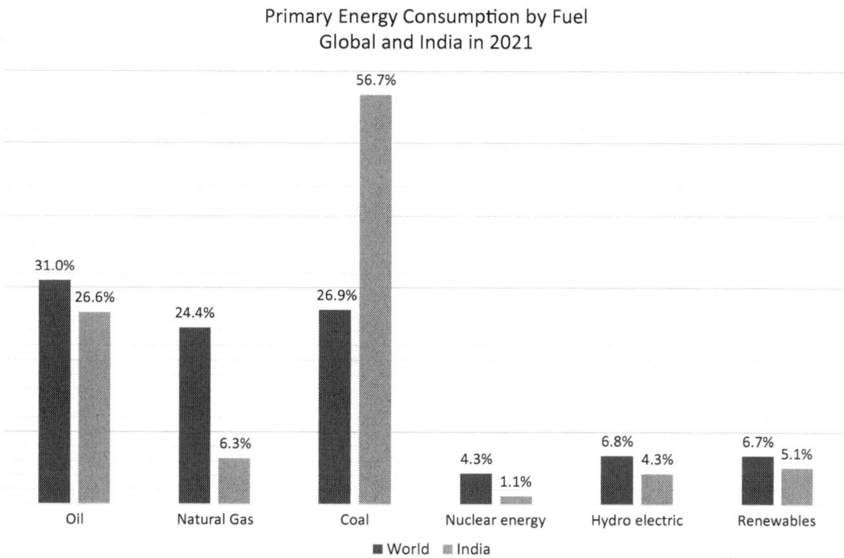

Figure 3.2 Consumption by fuel in the global and Indian primary energy mix in 2021.
Source: BP Statistical Review (2023)

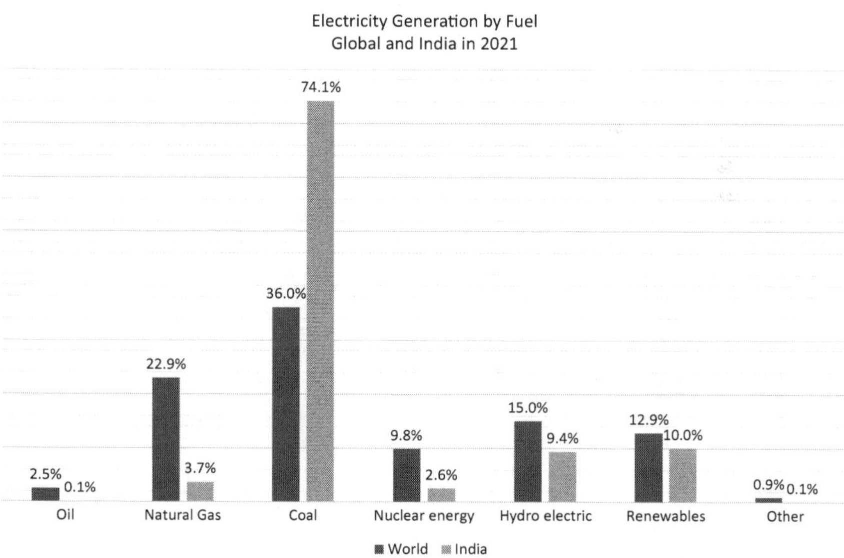

Figure 3.3 Electricity fuel sources globally and in India in 2021.
Source: BP Statistical Review (2023)

As per the latest available data of the CAIT dataset, GHG emissions in India stood at 3.36 Gt of carbon dioxide (CO_2)-equivalent in 2019, contributing 6.9 per cent to global GHG emissions (Climate Watch, 2021).

3.2.2 *The Climate Change Challenge and the Role of Natural Gas to Net-Zero*

India submitted its NDCs to the UNFCCC and ratified the 2015 Paris Agreement by making a number of pledges along with the reduction in the emissions intensity of GDP, the increase of non-fossil energy sources in installed electricity capacity, and the expansion of carbon sinks by 2030. The entire NDC document and the revised targets of 2021–2022 contain no single mention of a gas-based economy, which underlines the lack of a clear understanding of the role of natural gas in the Indian energy mix, not to mention a national road map. However, in February 2021, Prime Minister Modi reiterated the goal of a gas-based economy in his inaugural speech to the World Sustainable Development Summit, where he spoke about the allocation of US$60 billion for the development of domestic gas infrastructure, the ongoing expansion of the city gas distribution network, and an additional 100 districts to be connected to the network within the next three years (Modi, 2021).

As concerns GHG emissions, various studies found that natural gas is emitting up to 50 per cent less CO_2 compared to coal which admits the fuel to have some relevance in a climate-constrained energy future (Fulton et al., 2011; IEA, 2020). Also, with primary energy consumption of natural gas at 24 per cent globally, India still has a promising headroom for growth in gas usage even when a long-term vision for the sector must pay growing attention to the role of carbon capture, utilisation, and storage technology; biogas; and green hydrogen (IEA, 2021; BP, 2023).

At the 26th Conference of the Parties summit at Glasgow, Prime Minister Sh. Narendra Modi pledged to cut India's GHG emissions to net-zero by 2070, showcasing India's commitment to tackling climate change. Promises also include getting 50 per cent of India's energy requirement from renewable energy by 2030 and reducing carbon emissions by 1 billion tonnes until 2030. This essentially hints at two points: first, given the rising energy demand, natural gas as a fuel will continue to be in demand, and, second, natural gas will serve as a complementary fuel to support the growth of renewables. However, given the backdrop of COVID-19, along with today's energy markets, policies, technologies, and choices, there can be three possible energy outlooks as per International Energy Agency (IEA) (2021).

1. The Stated Policies Scenario (STEPS) – It assumes that pandemic would be brought under control by 2021, and existing policies and targets would remain as is, which is to say that practical constraints in implementation of such policies may limit the implementation in practice. Average growth in real GDP is expected to be 5.4 per cent between 2019 and 2040.

2. The India Vision Case – This scenario takes a more optimistic view on the pace of economic recovery and long-term growth along with full implementation of stated policies towards achieving a cleaner climate. Average growth in real GDP is expected to be 6 per cent between 2019 and 2040.
3. The Sustainable Development Scenario (SDS) – This scenario takes a goal seeking approach, working backwards from climate commitments, energy access goals including COP commitments and determining the platter of actions that would need to be taken today (IEA, 2021).

Fossil fuels account for less than 60 per cent in the SDS, compared with approximately 70 per cent in the STEPS and India Vision Case Scenario, while the current contribution of fossil fuels in the Indian energy mix stands at 90 per cent. The total demand of primary energy is 28 per cent less in SDS vis-à-vis STEPS.

In addition to this, geopolitical developments, such as Russia–Ukraine war, have further triggered the urgency to ensure energy security, with Europe primarily switching fuels and improving energy efficiency measures (FT, 2023). Given India's high dependence on natural gas imports, there is increased focus on Exploration and Production (E&P) activity. At the same time, the Green Hydrogen Mission has been launched to accelerate growth of green hydrogen production and consumption, achieving 5 Metric Million Tonne per Annum (MMTPA) domestic production by 2030. However, the natural gas infrastructure can still play a key role in lowering transportation cost by repurposing existing natural gas pipelines. The allowable percentage of blending would be subject to the steel used, current pipeline conditions, and investment required for minor modifications. Current desktop studies suggest that a 5–10 per cent blending should be possible with minor modification costs for India's natural gas pipelines.

3.2.3 Decarbonisation Pathways of Major Economic Regions

The IEA India Energy Outlook 2021 has captured energy-related CO_2 emissions relative to GDP per capita in major global regions for the time span of 1990 to 2040. Europe in particular shows that decarbonising the energy system is possible without compromising economic growth (see Figure 3.4). The dotted lines set in 2019 indicate an SDS leading into a net zero-emissions future across all regions without compromising solid economic growth and the achievement of the sustainable development goals.

The following section will introduce the Indian natural gas sector and discuss its key characteristics.

3.3 Status of the Gas Sector in India

The future of natural gas in India's energy mix is not fully outlined yet. In the most optimistic IEA scenario, natural gas could almost double by 2040, with applications in the industrial sector, through city gas distribution, and

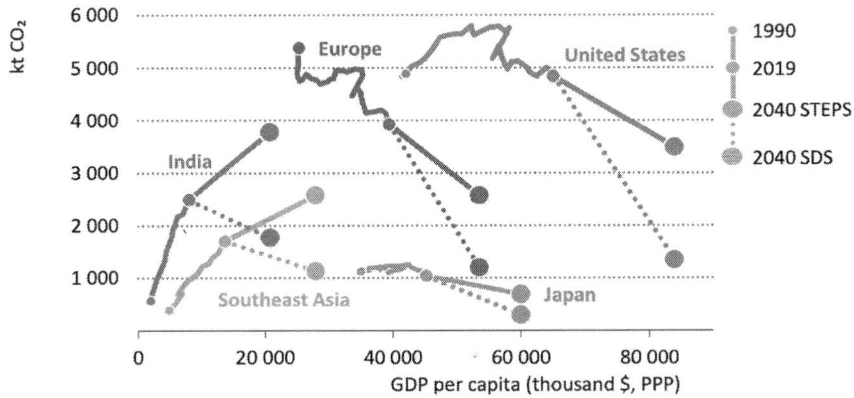

Figure 3.4 Energy-related CO_2 emissions and GDP per capita by region, 1990–2019–2040.

Source: IEA (2021). India Energy Outlook. URL: www.iea.org/reports/india-energy-outlook-2021 Licence: CC BY 4.0

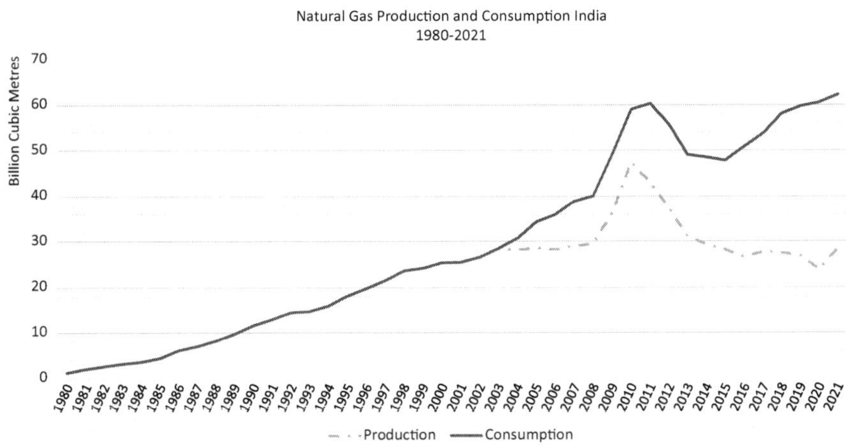

Figure 3.5 Natural gas production and consumption in India, 1980–2021.

Source: BP Statistical Review (2023)

by displacing coal electrification (IEA, 2021). The current perspective on gas usage from within the industry, however, is blurred by multiple structural, policy, and regulatory impediments, bringing the share of natural gas down to merely 6 per cent. Domestic production peaked in 2010 at 47.4 bcm, while consumption has grown to 62.2 bcm in 2021, as illustrated in Figure 3.5. Natural gas in industrial use has increased from 2 per cent in 2010 to 10 per cent in 2019. Similarly, the use of natural gas in the building sector has tripled over the same period (IEA, 2021). Since 2013, the gap between domestic

Table 3.1 Structure of Indian oil and gas sector administration.

Petroleum and natural gas sector		
Federal level	*Regulators/administrators*	*Other government agencies*
MoPNG	Upstream – DGH Midstream and Downstream – PNGRB	Oil Industry Development Board Oil Industry Safety Board Petroleum Planning and Analysis Cell NITI Aayog

production and consumption is ever-widening. In fact, domestic production is down to 2005 levels serving roughly 46 per cent of national consumption, despite proven gas reserves of 1.3 trillion cubic meters in 2019 (BP, 2021).

India currently imports around 33 bcm liquefied natural gas (LNG), largely from Qatar, Angola, Nigeria, the United Arab Emirates, and the United States, while in all probability continuing to be highly reliant on imports (BP, 2021). Whether domestic production could increase, imports grow, or both, multiple interventions would be required for increasing the share of gas to a targeted level of 15 per cent by 2030. The focus needs to be on developing a systematic road map for the usage of gas. But before we delve into the details of such a road map, we have to address key attributes of the Indian natural gas sector, including the evolution of its organisational and regulatory set-up, infrastructure, and market conditions.

3.3.1 Administrative Organisation and Policy Framework of the Gas Sector

The oil and gas sector business is divided into upstream (exploration and production), midstream (storage and transmission infrastructure), and downstream (refining, marketing, and distribution) operations. In India, the administration of the sector lies in the hands of the Ministry of Petroleum and Natural Gas (MoPNG). MoPNG is mandated to develop the sector, including formulating key policies to govern and monitor its operations. The office of the Director General of Hydrocarbons (DGH) assists the ministry in regulating the upstream sector, including calling competitive bids. As an administrative arm of the ministry, DGH also monitors the upstream sector. For an overview of the structure of the Indian oil and gas sector administration, see Table 3.1. The Petroleum and Natural Gas Regulatory Board (PNGRB), formed by an Act of Parliament in 2006, is an independent policymaker and regulator of the midstream and downstream sectors. NITI Aayog – the erstwhile Planning Commission – assists the sector in directing and planning long-term strategies. The Petroleum Planning and Analysis Cell is involved in day-to-day monitoring and administration. The Oil and Safety Directorate is a technical arm of MoPNG and assists the ministry in formulating safety rules and conducting audits.

The main Acts and rules governing the oil and gas sector include the Petroleum Act of 1934, the Oil Fields (Regulation & Development) Act of 1948, the 1959 Petroleum and Natural Gas Rules, and the Petroleum and Natural Gas Regulatory Board Act of 2006. In addition, there are multiple strategy and policy documents that shape the industry. They include:

- Indian Hydrocarbon Vision 2025
- Pipeline Policy
- National Exploration and Licensing Policy (NELP)
- Hydrocarbon Exploration and Licensing Policy (HELP)
- Policies for high pressure, high-temperature fields, and the like

In the following section, we elaborate on the role of these policies vis-à-vis infrastructure development.

3.3.2 Infrastructure

The development of infrastructure in any country depends upon a number of factors, including policy initiatives taken by the federal government to attract risk capital, the internal consistency of these policies, and the presence or absence of bureaucratic red tape. In India, key policy reforms started in the year 1997 with the introduction of the NELP round of bidding for new upstream acreages. This was internationally competitive bidding targeted to attract risk capital as well as technology to augment exploration and production of oil and gas in India. However, even after multiple rounds of bidding under NELP, participation from large oil and gas conglomerates remained mute. Insufficient investment and interest from large players keep plaguing the upstream sector. Out of the 254 blocks awarded under NELP, only 46 blocks are operational, whereas 206 blocks have been abandoned for various reasons. Large multinational players have given the sector a miss. Out of the 117 companies working in the Indian oil and gas sector, 48 are foreign companies. This resulted in NELP being changed to HELP in the year 2018. The HELP offered uniform licensing for all types of hydrocarbons, brought in open acreage (investors are free to bid anytime during the year based on the available central database of acreages), introduced a *revenue model* replacing the *cost-plus model* to decrease the monitoring costs in overall administrative expenditures, and provided for marketing freedom of gas produced from high pressure and high-temperature fields. Under the Open Acreage Licensing Policy (OALP) of HELP, out of 147 awarded blocks, 144 are operational (PPAC, 2020). However, more than the policy, the lack of large acreages at reasonable depths for exploration has attracted little interest from large players. Similarly, in the midstream and downstream sectors, participation remains restrained, except for some large domestic players investing in city gas licences.

The backbone of a gas transportation system is a country-wide pipeline network. In 2020, the total length of operational gas transportation pipelines amounted to 16.575 km, 88 per cent of which is controlled by government-owned companies. An additional 15,000 pipeline kilometres are under various stages of construction. PNGRB oversees the network expansion and regulates the tariffs for users of that infrastructure. However, even after the completion of these infrastructure projects, the pipeline density in the country remains inadequate, with many regions remaining unconnected. Comparatively more successful has been the setting up of import and regasification terminals for LNG. India currently has six LNG import terminals primarily at the west coast of the country (Gujarat, Maharashtra, Kerala, and Tamil Nadu on the east coast) and several under development in West Bengal, Odisha, Maharashtra, and Gujarat. The Regasified-Liquified Natural Gas (R-LNG) terminals are under the Open General Licence category which means that anyone can file an application for setting up an LNG import terminal. The current import capacity of existing terminals is 42.5 MMTPA. Another 20–30 MMTPA is under consideration or in the project phase. Constructing these terminals is important because the Indian oil and gas acreages are not very promising, thus necessitating the import from countries where gas is available in abundance like the Middle East, Africa, the United States, and Australia (PPAC, 2020).

At the city level, the issuing of licences for the building of city gas networks would broaden its applications in buildings, industry, power generation, and other commercial or business contexts such as Compressed Natural Gas (CNG) stations for mobility purposes. Currently, a total of 2,208 CNG stations exist across India and many more are planned in the coming months and years. The Government of India is also pushing for LNG stations across the country's main highways to serve heavy-duty road vehicles, ultimately contributing to lower local and atmospheric emissions.

3.3.3 Market

Gas markets in India can be described as oligopolistic, controlled by a few large players both private and public. Since the market is underdeveloped and competition limited, the government and regulators are intervening to restrict the regional monopoly. In the area of domestic oil and gas production, players were very few, even after the opening up of the downstream sector in the year 1997–1998. Transport infrastructure which forms the core and connects producers with customers is yet inadequate and a few major players monopolise the market by offering bundled services. Most of the domestic gas is allocated by the federal government as per the priority of the gas utilisation policy. The price is determined every six months based on a formula that serves as a proxy for the market value of gas.

Affordability is a sensitive issue for consumers. The key customer segments which can afford gas include liquid fuel replacement customers in the refining, petrochemical, fertiliser, and the mobility sectors. Electricity generation from gas, which is an anchor load for gas production and transportation across the world, however, makes little economic sense in India because it is comparatively more expensive. The merit order principle renders electricity generated from gas non-dispatchable. The electricity distribution market is further distorted by tariffs which as a people-friendly measure are kept artificially low by political parties at the state level. As a result of all this, coal keeps dominating the electricity sector.

3.4 The Evolution of the Indian Gas Market

It is imperative to comprehend the evolution of any sector to learn from experience, make corrections where needed, and move on. We can divide the evolution of the Indian oil and gas sector into four distinctive phases: (1) The Nomination phase (until 1990), (2) the pre-NELP phase (1991–1997), (3) the NELP phase (1998–2014), and (4) the HELP phase starting from 2015 (Mehrotra, 2017).

3.4.1 Government Control or Nomination Phase (Until 1990)

In the initial stages of the development of a country, it is difficult to attract capital on a competitive basis. Most governments either end up spending capital on their own or nominate experienced companies to start investments. In this initial phase, the risk-reward is in favour of the investor.

During this period, the Government of India provided acreages on a nomination basis mostly to public sector companies. The gas price was determined by the cost-plus principle under the Administered Price Mechanism. To demonstrate the viability of the midstream segment, the first interstate trunk gas pipeline, Hazira–Vijaipur–Jagdishpur, was set up by the Gas Authority of India – a government-owned company established in 1984. The pipeline became operational in 1991, with an anchor load from fertiliser plants in Uttar Pradesh.

3.4.2 Negotiated Corporate Monopoly – the pre-NELP Phase (1991–1997)

During this second phase, the government needed more technological know-how, along with risk capital. Consequently, 100 per cent foreign direct investment (FDI) was allowed for exploration and production, and the discovered Panna Mukta and Tapti fields in the Mumbai High Complex were awarded to international oil companies in the year 1994. Production sharing contracts (PSCs) were signed stipulating the cost-sharing principle and the formula for the price of gas. The Indian government wanted an independent institution to manage the sector which resulted in the set-up of the office of the DGH in the year 1993. In gas distribution, the government invited national and international companies to set up city gas networks in Mumbai and Delhi under the FDI limit of 50 per cent.

3.4.3 Early Market Liberalisation – the NELP Phase (1998–2014)

India started its economic liberalisation in the year 1991, but the oil and gas sector had to wait until 1999 when formulated NELP policy took effect in 1997.

For the upstream sector, 100 per cent FDI had been allowed in the pre-NELP era. In the midstream and downstream sectors, however, state-owned companies continued their monopolistic businesses for much longer. Only with the ratification of the Petroleum and Natural Gas Regulatory Board (PNGRB) Act in 2006 and the set-up of the Regulatory Board – the first independent regulator for the oil and gas sector – the market did start to liberalise. In the next step, 100 per cent FDI was allowed into the midstream and downstream sectors. PNGRB issued regulations to independently manage the sector and introduced competitive bidding at the international level for the development of pipelines and city gas networks.

However, this phase of the liberalisation process experienced many challenges, including a huge number of litigations under the cost-sharing mechanism of the PSC. The PSC had envisaged giving pricing freedom to the contractor which never materialised. Instead, the government constituted various committees and introduced a proxy formula to mimic competition. The Rangarajan approach to gas pricing assumed that there would be a continued lack of a free market to determine a competitive price for domestic gas. Hence, the policy for gas pricing would have to be based on cues from global competitive gas trading (GoI, 2012). The currently used formula, which was devised by Dr Vijay Kelkar, recommends gas pricing to be determined through volume-weighted linkages to international hubs such as National Balancing Point (UK), Henry Hub (USA), Alberta (Canada), and Russian Hubs (Kelkar, 2014)

The midstream and downstream bids for pipeline and city gas network development were also highly manipulated. The bidding entities gained licences through very low to near zero tariffs, while committing huge work programmes. As a result, many of these pipelines and networks never materialised.

3.4.4 Introduction of Partial Competition – HELP Phase (2015 and Beyond)

The last phase in any market evolution envisions the introduction of full competition where all decisions are made by market forces or by an independent regulator in the case of a monopoly business. However, in the Indian context, this remained mere intention. The Indian Government did not have the political will to fully open up and deregulate the market. The years after

2015 remained a phase of partial competition even if the intention to move towards the full opening of the gas market was still entertained. Nevertheless, the last couple of years have seen major decisions, some of which are more regressive in nature.

- The introduction of HELP policy changed the PSC regime from cost-sharing to revenue-sharing. The clear positives from the HELP policy include a single licence for conventional and non-conventional hydrocarbons along with OALP. High pressure and high-temperature acreage, however, keep the debate alive whether HELP policy is the most suitable.
- With the introduction of OALP, a far-reaching reform has been initiated, which will allow companies to determine the timing of licensing.
- Pricing freedom for yet-to-be harvested fields (as of January 1, 2016) including deepwater, ultra-deepwater, and high-pressure high-temperature fields. However, after giving price freedom, the government has put a price ceiling in place. The price ceiling is determined by the lowest of (1) landed cost of imported fuel; (2) weighted average of landed cost of imported substitute fuel or (3) landed cost of imported LNG.

In addition, the regulator has finalised a unified tariff for transmission pipelines, which will be conducive to distributing the gas across the country more equally.

3.5 Challenges in Building a Gas-Based Economy

India is a unique and diverse country in many respects. It is a young and ambitious, newly industrialising country, with a rich and deep history, confronted with multiple developmental challenges at the same time. The development of its natural gas sector can only give us at best a glimpse at the big picture of these much larger challenges. Some of the main issues in India's gas sector development are discussed below distinguishing policy, regulatory, fiscal, and financing matters and their effects upon operations.

3.5.1 Policy, Regulatory, and Fiscal Issues

3.5.1.1 Precipitate Policy Activism

In the past, Indian policymakers have infamously changed policies related to the development of the gas sector multiple times. This has left industrial players with seeing the goalpost being moved to various places and positions. Industrial players had to hear policy being discussed without any principled approach or rule-based mechanism that would have established a robust framework for operations over the next 10–15 years. By contrast, policy changes oftentimes took retrospective effect, changing the whole viability and return on the investment potential of a project. Large sums of capital are waiting to be invested in India from both domestic and foreign players.

However, uncertainties related to policy and policymaking frequency carry their mark on evaluating any project on a long-term basis. Moreover, the regulatory activism is infused by ideological distortions, depicting profit-making as a cardinal sin which Indian policymakers tend to fight with the means of tax warriors – an attitude that has effectively dissuaded risk capital.

3.5.1.2 Fiscal Regime

There is a need for bringing the full supply chain (producer, regasification terminal, transporter, and end-consumer) under the same tax regime in order to optimise the cost. The biggest tax-related downside comes from gas not being included under the national Goods and Services Tax (GST) regime allowing states to tax gas at higher rates. This makes gas substantially costlier compared to other fuels because the state tax on gas cannot be offset from the end product even if the end product is covered by GST.

3.5.1.3 Complex Tariff Structure

A complex tariff structure of additional charges and tariffs has led to segmented market growth. A multiple tariff structure renders gas more expensive for users more distant from the source. The IEA (2021) reckons with, on average, double the cost of wholesale gas for the final user in the year 2019. However, with a unified tariff structure notified by PNGRB, this cost impediment would soon be addressed.

3.5.1.4 No Pricing and Marketing Freedom

Gas is produced from the fields falling under different regimes as outlined in Section 3.4, that is, the fields allocated in the pre-NELP, NELP, and HELP phases. For each of these phases in the evolution of the Indian Gas market, the government has allowed different price points. In order to expand production, HELP, for example, allows for pricing and market freedom for gas produced from deepwater, ultra-deepwater, and other complex reservoirs. For all other fields, until a market mechanism takes shape, the price is determined on a six-month basis by the Kelkar formula linked to international hub prices. The comparatively low gas price of the past decade has further discouraged notable investments in domestic production, and because of an illiquid Indian gas market with meagre to no trade volumes, free price discovery does not happen. The development of an Indian gas hub is the right step in that direction to provide price signals towards gas trading and development of gas markets.

3.5.1.5 No Balance in Risk-Reward

For the developer, who carries the full bandwidth of risks, current regulation caps post-tax returns at 12 per cent. Given the capital intensity of projects, the bottlenecks in the supply of gas, the risk-reward becomes a largely unattractive proposition for investors.

3.5.1.6 Distortions in the Energy Market

Gas has not received a fair chance to be the fuel of choice in India, due to multiple distortions in the energy sector. This applies first and foremost to the electricity sector, in which tariffs are highly subsidised by each state government and a preference for new renewable energies effectively displaces the chances of natural gas. Over the years, the use of natural gas for power generation has decreased due to extreme financial pressure and operating on very low capacities, preventing gas from finding a viable place in the merit order dispatch.

3.5.1.7 Multiple Planning Institutions

The lack of integrated planning is a bane for many sectors. In India, the planning is split between NITI Aayog and the MoPNG, while both act out of sync. Separate ministries for coal, power, petroleum, and renewables further compound this problem with each ministry pulling policy formulation in different directions to meet their own narrow objectives that are often at crossroads with the overarching national goals.

3.5.1.8 Misaligned Federal and State Governments – Multiple Window Clearance

Misalignment between centre and state is not unique to the Indian federal structure; however, it is adversely affecting the development of the gas sector, especially midstream and downstream activities. States still control the land and labour laws and it is agonisingly difficult for companies to secure the right of way for laying pipelines and city gas networks in towns and districts. Coupled with this is the multiple clearances requirement. Developers have to get "Green Clearance" from authorities to cross forest land, to get right of way through railway crossings, river crossings, and highways, which can lead to years of delay in implementing any project.

3.5.1.9 Monopolistic Positions of Incumbents and Lack of Transparency

There are only a few players covering the entire value chain in the Indian energy sector. Upstream, which requires the largest capital, has failed to attract oil majors. The scenario in midstream is even worse, where there are mostly government players and bundled offerings limit access to pipelines. Capacity booking in pipelines, for example, is at the sole discretion of the operator and still an unnecessarily complicated process. Downstream has seen some competition in the city gas space but has also been consolidating with a few large players. Due to a few large players controlling the market and inadequate reporting requirements, a lack of transparency enshrouds the sector and further impedes the development of markets. Overall, the same actor plays multiple roles many times, that of a gas marketeer, the transporter, capacity manager, and the like. There is a need for developing new players with a different set of roles, for example, in capacity management.

3.5.1.10 Lack of Concrete Push From Central Government

For the past two decades, the central government has made a lot of noise to increase the share of gas from 6 per cent to 15 per cent in the energy mix. However, no substantial policy initiatives have been taken to push the cause of gas. In contrast, concrete goals have been formulated and tangible achievements made to increase the share of renewables in India to 175 GW by 2022 and 450 GW by 2030. This raises questions about the seriousness with which the Indian government is presenting its ambitions for a larger share of gas being used in the Indian economy.

3.5.2 Financing

Infrastructure projects require a huge capital outlay and years of gestation period. Hence, it becomes important for Indian policymakers to provide suitable financing options to attract investments. In recent years, the advantages created by the Investment Infrastructure Trust (InvIT) have attracted attention, but there is still a need to simplify its structure and to broaden the financing options for gas infrastructure development. Restrictions on what capital can be invested in InvITs weaken the instrument. Currently, a huge amount of capital is lying with insurance and pension funds (IFRDA/PFRDA), which policy restrictions prevent from investing in InvITs. Investors, on the other hand, are wary of investing through such structures because they are currently given only limited control over their funds. This in turn results in a bottleneck in both running and optimising the investment channel efficiently.

3.5.3 Overall View: Putting the Cart Before the Horse

The pace at which any reform has to be implemented needs prior research and deliberation. The government and regulator went about the partial opening of the gas market with very basic infrastructure. Attracting capital for the infrastructure sector in the initial phase needs a higher premium to the risk-free return. However, with the opening up of the market, a wrong signal was sent: the returns would be competitive. This led to suboptimal capital outlay from the private sector. Most markets open up only after a substantial part of the infrastructure is in place, which has not been the case in India.

PNGRB, the midstream, and the downstream regulators were quick in establishing various rules and regulations in the first three years of their operations. This early and hastened regulatory push meant that most regulations were not mature enough to stand the test of time. In other cases, PNGRB has failed to implement its own regulations. A case in point is the access code regulation, where missing enforcement led to the blocking of access to pipelines and networks, bundled products, and opaque companies.

In summary, the gas markets in India though opened up for international competitive bidding and FDI of up to 100 per cent have not witnessed the growth envisioned by the government. Today, there are only a limited number of large companies that are exploring the Indian basins, as a result of which the infrastructure of a high-pressure cross-country pipeline network is grossly underdeveloped, and there are no aggregators and shippers of natural gas other than government enterprises. At the same time, there are no easy answers to the above-listed challenges.

Henceforth, the Indian gas markets need to be recalibrated to overcome what reminds in a catch-22 situation. The government should do so by first making natural gas available and accessible with limited competition. Only in a later step, full-scale competition should be introduced to make it more affordable. As seen across various geographies, financial markets will shadow the physical market conditions. International gas markets have gone through this process and there is something to learn from their experiences.

3.6 Insights From Advanced International Gas Markets

There are many countries with flourishing gas markets. We have selected two matured markets – the United States and the United Kingdom – to provide some insights into international best practices and lessons learned that can be meaningful for the Indian context.

3.6.1 US Market

The US gas market is perhaps the most mature with a history of over 70 years. The development of the US market started with the 1938 Gas Act. The Federal Energy Regulatory Commission (FERC) is the regulatory body empowered by the Gas Act to organise the sector and the companies involved in interstate transactions. Until the year 1954 – for more than 15 years since the Gas Act – the United States had no wellhead price regulation. This meant that upstream companies were free to marketise and price their gas. This era, which was also backed by the post-World War economic boom, saw massive investment in both upstream and development of interstate pipeline networks. However, FERC got powers to regulate wellhead prices as the result of the US Supreme Court ruling deciding over complaints of price rigging (Juris, 1998). This order was implemented in 1954 and led to a huge rise in demand for gas because prices were set artificially low (even lower than the unregulated coal and oil). This resulted in multiple distortions in the US market. By the year 1970, slower investment in upstream led to lower revenues and gas shortages. The high demand for gas spurred the faster development of interstate pipelines. Local distribution companies had to sign long-term take or pay contracts to secure capacity, which resulted in long-term liabilities (Herbert & Kreil, 1996; Pierce, 2004). The 1978 Gas Act corrected these problematic entanglements by removing the price ceiling on wellhead gas which re-attracted investments in the upstream sector. In the late 1970s and 1980s, the first spot market got created in the United States. To further correct many market anomalies, the US regulator came out with a series of reforms (GAO, 2006; Talus, 2014):

- FERC Order 380 relieved local distribution companies from taking or paying contracts with pipeline operators.
- FERC Order 436 mandated pipeline operators to provide open access voluntarily.
- The 1989 Natural Gas Wellhead Control Act removed multiple price points prevalent in the US market by ultimately removing wellhead price regulations. The Act also unbundled the transporting and marketing activity on interstate pipelines helping to open access and free movement of gas.
- FERC order 636 further strengthened open access by enforcing the unbundling of pipeline ownership.

These far-reaching measures led to the creation of hubs (market centres) for the trading of gas at various pipeline intersection points. FERC also allowed trading in secondary capacity and capacity release provisions. Over the years, FERC took multiple initiatives to strive for transparency and continued reforming the sector. Import terminals, for example, were allowed to charge market rates and to restrict access if set up on a merchant basis. This resulted in large investments made in terminals. However, with the US shale gas boom, these import terminals were converted into export terminals serving their purpose in a global gas economy (De Vany & Walls, 1994; Pierce, 2004; Talus, 2014).

3.6.2 UK Market

Similar to the US market, the British gas (BG) market has many years of experience starting in 1948. One could say that the United Kingdom saw the wheel turn full circle during its rich history. Before the year 1948, the United Kingdom had 1,046 private companies supplying town gas (which was syngas from coal) to the various councils. The Gas Act of 1948 nationalised the UK gas industry by merging all these companies into 12 area councils or gas boards, which created a vertically integrated monopoly. The first gas discovery in the United Kingdom took place in the year 1966. A high-pressure transmission network was created to connect the 12 area councils and town gas was replaced by natural gas. The United Kingdom was the only country in Europe to import LNG from Nigeria during this period. However, there was no gas pipeline connection between the United Kingdom and continental

Europe at that time (Stern, 1997; Calliope, 2009; Thuber & Chang, 2011). The integration was further strengthened with the passing of the 1972 Gas Act, through which the British Gas Corporation was formed by merging all the 12 area councils. BG was given monopsony and monopoly powers. It bought gas on negotiated terms with wellhead price freedom but undersupply specifications and sold gas on a cost-plus basis. After creating a monopoly business that was conducive to infrastructure development as well, the United Kingdom commenced opening up of markets in the 1980s (forced by stagflation in 1978) through a series of reforms (Simmonds & Bartle, 2004; EFET, 2004; Calliope, 2009).

- The Oil and Gas Enterprise Act of 1982 removed pre-emption rights of BG on purchase of gas.
- The 1986 Gas Act formed the independent Office of Gas Regulation (OFGAS). BG was privatised forming BG Plc. Large customers were given the choice to buy gas using open access to pipes.
- A monopolies commissioner barred BG Plc in 1988 from buying 90 per cent of new gas. BG was asked to publish tariffs.
- The Gas Release Programme of 1992 asked BG to reduce its market share to 40 per cent in three years. The threshold for open access was reduced. BG was reorganised into three companies for upstream, midstream, and downstream.
- The Gas Act of 1995 introduced competition in the residential market and unbundled transportation from the supply. Spot markets evolved due to oversupply.
- The Access Code of 1996 addressed all elements of open access.
- BG was legally unbundled into BG Plc (upstream) and Centrica Plc (gas trading and retailing). The full competition was introduced in residential markets.

In the late 1990s, the first pipeline connecting the United Kingdom and continental Europe was completed. A National Balancing Point was established in the year 2000, and the UK government merged gas and electricity regulators to form the Office of Gas and Electricity Markets (Calliope, 2009).

The Learnings from both US and UK markets have many similarities. However, each country is unique, and India as a country has many unique features as well, including price sensitivity and a reluctance to pay for cleaner energy. One independent agency to govern the sector would be needed to curate the learnings for Indian conditions as summarised as follows:

- Consistent policy reforms starting with allowing monopoly power to help development of infrastructure and gradual opening up of the markets.
- Unbundling of content and infrastructure to allow open access without any conflict of interest.

- Development of physical markets with multiple market players followed by financial markets.
- Ensuring energy security for the country while deciding the reform trajectory.

3.7 Road Map for a Gas-Based Economy

On the basis of the learnings from developed gas markets, the hitherto evolution of the Indian gas sector, and keeping in view its challenges, we recommend a five-step road map for the development of a gas-based Indian economy. The democratic foundations of India render wholesale changes unrealistic and would meet with deep-seated political resistance. Accordingly, we propose the following phased approach:

Stage 1 – Getting the Foundations Right

The stability of policy, fiscal regime, and rules governing a sector are the cornerstone of attracting investments. Unstable, changing, or not implemented policies develop an atmosphere of mistrust and lead to unnecessarily time-consuming litigations between government and private investors. The government should also invest in bringing various planning institutions together in order to have a comprehensive and shared understanding of the role each source of energy is going to play in the future energy mix of the country. As concerns the gas sector, developing regulatory institutions that are competent and independent will go a long way in developing the sector. Based on Mehrotra and Gupta (2020), the following are the key recommendations for the government at this stage:

- Outlining a clear energy policy defining the role of each fuel including the role of gas in complementing renewable energies.
- Articulating and building confidence that policies would be stable and policy changes would apply only prospectively.
- Including the entire gas value chain under GST to create a level-playing field.
- Developing competencies in the regulatory board by making them financially independent and allowing recruitment on a merit basis instead of a conflict-prone deputation from the public sector.
- Converting the office of the DGH into an independent regulatory entity.

Stage 2 – Focus on Infrastructure Development

Development of markets would first need gas inadequate quantity both locally and from imported sources. Since the Indian gas market is highly price-sensitive and imported sources see price fluctuations that are controlled by international demand and supply, it is imperative to develop large indigenous resources. Also, gas as a commodity is unique in that it needs a network of cross-country pipelines to reach customers. This would require increasing the length of pipelines by at least three to four times to make them available across the country. The following are the key recommendations for this stage:

- Providing production sharing contracts, balancing risk, and return. (A revenue sharing mechanism may be too risky for high-pressure high-temperature frontier acreages.)
- Providing marketing and pricing freedom to all producing fields.
- Attracting investments by changing vexed regulatory issues such as bidding norms for pipelines.
- Providing single-window clearance for the right of way for laying cross-country pipelines and city gas pipeline network.
- Using InvIT structures to monetise pipelines held by the government and using the proceeds to develop pipelines in areas that are suboptimal for the private sector.

Stage 3 – Start Liberalisation

Liberalisation is required not only in the gas sector but in the energy sector as a whole for gas to play its part in the energy mix. The government needs to bring in a level-playing field based on carbon neutrality for each energy source to play its adequate part. The key measures to be taken during this phase are as follows:

- Removing subsidies from power sector tariffs and instead of providing direct subsidies to the poor.
- Introducing a carbon tax on the pollution generated by fossil fuels.
- Removing the right of government companies to act as an aggregator of gas for the fertiliser and power sector.
- Formulating more stable open access norms for free movement of gas.
- Changing the transmission tariff regime to a unified tariff for gas to reach all corners of the country.
- Removing all restrictions on wellhead pricing and marketing and allowing all fields to be treated under one norm.

Stage 4 – Break Monopoly Power

Monopoly powers come with vested interests of the government which arguably has an instinct to retain control over the sector. However, the sector would benefit from the following actions:

- Unbundling transportation and marketing of gas and if required implementing legal separation. Bifurcating GAIL businesses into three parts, that is transportation, supply, and distribution of city gas.
- Relieving the market from its take or pay contracts with incumbents; asking for new contracts which establish a level-playing field.
- Mandating transmission and city gas companies to be transparent in their operations through public bulletin boards, for example.
- Setting up a central gas management agency for open access and transmission of gas.
- Enabling a more functional gas exchange.

Stage 5 – Full-scale Liberalisation

Full liberalisation is a continuous process rather than a steady state. Interventions from the government are required on a rolling basis, innovating as per the need of the sector. Full liberalisation is nothing but increasing the breadth and depth of the market. The following three points speak about potential moves towards this end:

- Ensuring multiple increases in physical trading to generate liquidity.
- Developing financial products for trading and hedging.
- Focusing on energy security and developing gas storage.

3.8 Conclusion

India has the potential to increase the share of gas in its energy mix from currently 6 per cent to a robust 15 per cent. This would help develop a gas-based economy, which is not discordant with the climate objectives. However, this requires careful multi-stakeholder planning and the execution of multiple phased measures within the next one to three years. In an ideal scenario, the stages as described earlier would be implemented in sequence. However, as was the case with developing a COVID-19 vaccine, all stages mentioned earlier could and would also have to work in parallel. In fact, India is at a point where a few actions of each stage are already in progress and where different regions of the country are at different stages in the development of the gas sector. The North and West of India, for example, can quickly move to full liberalisation since many actions are already taken, while the rest will follow in the next few years. With that, this decade will see a gas-based economy emerging more strongly in India.

References

Benoit, P. (2019). *Energy and Development in a Changing World: A Framework for the 21st Century*. Center on Global Energy Policy, SIPA, Columbia University. Retrieved from: https://bit.ly/32F01jE (Last accessed: 24.04.2021)

BP (2021). *What is the Role of Gas in the Energy Transition?* British Petroleum (BP). Retrieved from: https://on.bp.com/32HxqKD (Last accessed: 24.04.2021)

BP (2023). *BP Statistical Review of World Energy 2023*, 71st edition. British Petroleum (BP). Retrieved from: www.bp.com/en/global/corporate/energy-economics/statistical-review-of-world-energy.html (Last accessed: 13.04.2023)

Calliope, W. (2009). *The Evolution of the Gas Industry in the UK. A Case Study Prepared for the International Gas Union's Gas Market Integration Task Force*. Retrieved from: https://bit.ly/3aBUxup (Last accessed: 24.04.2021)

Climate Watch (2021). *Historical GHG Emissions*. Retrieved from: www.climatewatchdata.org/ (Last accessed: 24.04.2021)

De Vany, A. & Walls, W.D. (1994). Natural gas industry transformation, competitive institutions and the role of regulation: Lessons from open access in US natural gas markets. *Energy Policy*, 22(9): 755–763. https://doi.org/10.1016/0301-4215(94)90051-5

EFET (2004). *EFET Principles for European Gas Market Development*. European Federation of Energy Traders (EFET), Amsterdam. Retrieved from: https://bit.ly/3sMTRII (Last accessed: 24.04.2021)

FT (2023). EU gas usage falls 18% after price shock caused by Russian supply cuts. *Financial Times*. 19.04.2023. Retrieved from: www.ft.com/content/a4d4e1ce-5444-42c7-bda6-6fda77cc2fe7?shareType=nongift (Last accessed: 26.04.2023)

Fulton, M., Mellequist, N., Kitasei, S. & Bluestein, J. (2011). *Comparing Life-Cycle Greenhouse Gas Emissions from Natural Gas and Coal*. Deutsche Bank Group, DB Climate Change Advisors, Frankfurt (M.). Retrieved from: www.westerngrid.net/wp-content/uploads/2012/08/Natural_Gas_LCA_Update_082511.pdf (Last accessed: 12.02.2021)

GAO (2006). *Natural Gas – Factors Affecting Prices and Potential Impact on Consumers*. U.S. Government Accountability Office (GAO). Retrieved from: www.gao.gov/products/gao-06-420t (Last accessed: 24.04.2021)

GoI (2012). *Report of the Committee on the PSC Mechanism in Petroleum Industry*. Government of India, December 2012. Social News XYZ. 06.02.2023. Retrieved from: https://taxindiaonline.com/RC2/pdfdocs/wnew/PSC_Report.pdf (Last accessed: 16.07.2021)

Gopi (2023). *India Committed to Energy Efficiencies, to Increase Hydrocarbon Production: Hardeep Puri*. Retrieved from: www.socialnews.xyz/2023/02/06/india-committed-to-energy-efficiencies-to-increase-hydrocarbon-production-hardeep-puri/ (Last accessed: 14.04.2023)

Gupta, A. (2019). India insight: $10 trillion GDP by 2030? Not quite, but almost. *Bloomberg Intelligence*. Retrieved from: https://bloom.bg/3ngNOLJ (Last accessed: 24.04.2021)

Herbert, J.H. & Kreil, E. (1996). US natural gas markets: How efficient are they? *Energy Policy*, 24(1): 1–5. https://doi.org/10.1016/0301-4215(95)00145-X

Howarth, N., Galeotti, M., Lanza, A. & Dubey, K. (2017). Economic development and energy consumption in the GCC: An international sectoral analysis. *Energy Transit*, 1(6). Springer. https://doi.org/10.1007/s41825-017-0006-3 (Last accessed: 24.04.2021)

IEA (2020). *Methane Tracker 2020*. International Energy Agency (IEA), Paris. March 2020. Retrieved from: www.iea.org/reports/methane-tracker-2020 (Last accessed: 19.04.2021)

IEA (2021). *India Energy Outlook 2021*. International Energy Agency (IEA), Paris. February 2021. Retrieved from: www.iea.org/reports/india-energy-outlook-2021 (Last accessed: 12.03.2021)

Juris, A. (1998). *Development of Competitive Natural Gas Markets in the United States. Viewpoint: Public Policy for the Private Sector; Note No. 141*. World Bank, Washington, DC. Retrieved from: https://openknowledge.worldbank.org/handle/10986/11557 (Last accessed: 23.04.2021)

Kelkar, V. (2014). *Roadmap for Reduction in Import Dependency in the Hydrocarbon Sector by 2030. Final Report*. Ministry of Petroleum and Natural Gas (MoPNG), Government of India. September 2014. Retrieved from: http://petroleum.nic.in/sites/default/files/FinalReportKelkarCommittee2014.pdf (Last accessed: 23.04.2021)

Mehrotra, A. (2017). Issues and challenges in development of efficient gas markets. In: Kar, S.K. & Gupta, A.K. (Eds.), *Natural Gas Markets in India*. Springer Nature, Singapore, pp. 197–215.

Mehrotra, A. & Gupta, A. (2020). Indian gas market – roadmap for creating an efficient gas market. In: Gupta, A. & Dalei, N.N. (Eds.), *Energy, Environment & Globalization – Recent Trends, Opportunities & Challenges in India*. Springer, Singapore.

Modi, N.D. (2021). *PM's Speech at the World Sustainable Development Summit*. 10.02.2021. Retrieved from: www.narendramodi.in/prime-minister-narendra-modis-speech-at-world-sustainable-development-summit-553860 (Last accessed: 12.02.2021)

O'Neill, A. (2021). *Global Gross Domestic Product (GDP) at Current Prices from 1985–2025*. Retrieved from: www.statista.com/statistics/268750/global-gross-domestic-product-gdp/ (Last accessed: 23.04.2021)

Our World in Data (2023). *Global Primary Energy Consumption by Source*. Retrieved from: file:///Users/abs/Desktop/artwork/global-primary-energy.svg (Last accessed: 27.11.2023)

PIB (2020). *PM Delivers Inaugural Address at 4th India Energy Forum*. Press Information Bureau (PIB), Government of India. Retrieved from: https://pib.gov.in/PressReleseDetailm.aspx?PRID=1667654 (Last accessed: 24.04.2021)

Pierce, R.J. (2004). Reconstituting the natural gas industry from wellhead to Burnertip. *Energy Law Journal*, 25: 57–112. Retrieved from: www.eba-net.org/assets/1/6/12_25EnergyLJ57(2004).pdf (Last accessed: 23.04.2021)

PPAC (2020). *Ready Reckoner. Oil Industry Information at a Glance*. Petroleum Planning & Analysis Cell (PPAC), Ministry of Petroleum & Natural Gas, Government of India. November 2020. Retrieved from: www.ppac.gov.in/WriteReadData/Reports/202101120159244134186PPACRRNov2020WebVersion.pdf (Last accessed: 23.04.2021)

PWC (2014). *Future of India. The Winning Leap*. PricewaterhouseCoopers India. Retrieved from: www.pwc.com/sg/en/publications/assets/future-of-india-the-winning-leap.pdf (Last accessed: 23.04.2021)

Ritchie, H. (2019). *Global Primary Energy Consumption by Source*. Our World in Data. Retrieved from: https://ourworldindata.org/energy-mix (Last accessed: 22.04.2021)

Sharma, P. (2020). India can be a $7 trillion economy by 2030. Exclusive interview with Arvind Panagariya, former Vice Chairman, NITI Aayog. Retrieved from: www.theweek.in/theweek/business/2020/03/26/india-can-be-a-7-trillion-economy-by-2030.html (Last accessed: 24.04.2021)

Smil, V. (2017). *Energy and Civilization: A History*. MIT Press, Cambridge, MA.

Simmonds, G. & Bartle, I. (2004). *The UK Gas Industry, 2003/2004*. Centre for the Study of Regulated Industries (CRI), CRI Industry Brief, University of Bath School of Management, Bath, UK.

Stern, J.P. (1997). The British gas market 10 years after privatisation: A model or a warning for the rest of Europe? *Energy Policy*, 25(4): 387–392. https://doi.org/10.1016/S0301-4215(97)00021-9

Stjepanović, S. (2018). Relationship between energy consumption and economic growth in 30 countries in Europe. *Croatian Economic Association*, 69(1): 43–57. February 2018. https://doi.org/10.32910/ep.69.1.3

Talus, K. (2014). United States natural gas markets, contracts and risks: What lessons for the European Union and Asia-Pacific natural gas markets? *Energy Policy*, 74: 28–34. https://doi.org/10.1016/j.enpol.2014.07.023

Thuber, M.C. & Chang, J. (2011). *The Policy Tightrope in Gas Producing Countries – Stimulating Domestic Demand without Discouraging Supply. Pacific Energy Summit, Held February 21–23, 2011, in Jakarta, Indonesia*. Retrieved from: www.nbr.org/wp-content/uploads/pdfs/programs/PES_2011_Thurber_Chang.pdf (Last accessed: 22.04.2021)

Timperley, J. (2019). *The Carbon Brief Profile: India*. Retrieved from: www.carbonbrief.org/the-carbon-brief-profile-india (Last accessed: 22.04.2021)

USAID (2018). *Greenhouse Gas Emissions in India. Climate Links Factsheet*. United States Agency for International Development. Retrieved from: https://bit.ly/3aNW8xx (Last accessed: 19.04.2021)

World Bank (n.d.). *World Bank Data*. Retrieved from: https://data.worldbank.org/country/IN (Last accessed: 18.04.2021)

4 Green Hydrogen

India's Billion-Dollar Opportunity

Tirtha Biswas and Deepak Yadav

4.1 Introduction

The year 2020 would be remembered for the COVID-19 pandemic that engulfed the world, leading to almost five million fatalities and a global economic slowdown. The year 2020 would also be remembered for an event that had to be postponed – The Tokyo Olympics – Japan's opportunity to host the first-ever "Hydrogen Olympics." However, with the receding pandemic and efforts to deliver an effective vaccine, the world finally witnessed the hydrogen flame on the Olympics torch in 2021.

While the COVID-19 pandemic is a short-term menace, climate change remains the long-term existential threat to the global community. At 21st Conference of the Parties in Paris, in December 2015, the parties to the United Nations Framework Convention on Climate Change reached a landmark agreement to keep the global temperature rise well below 2°C above pre-industrial levels. At 26th Conference of the Parties in Glasgow, India gifted the world with "Panchamrit" formula that included a net-zero pledge by 2070 and other 2030 targets. In this context, a potentially zero-carbon fuel like hydrogen provides an opportunity to help decarbonise major end-use sectors of the economy – various industrial processes, electricity storage, and the transport sector. However, the prospect of large-scale adoption of hydrogen must be taken with a grain of salt. In the past, the use of hydrogen as an energy carrier and fuel has seen multiple ups and downs. Today, it is again on the horizon as the global community perceives hydrogen generated from renewable sources of energy as critical to the deep decarbonisation of the global economy. However, unlike in the past, hydrogen seems to be the closest to reality. Many countries have announced hydrogen road maps targeting gigawatt (GW)-scale installations. Countries like Japan are already in the implementation phase with megawatt (MW)-sized electrolysers. Japan is also a potential export market for blue and green hydrogen and derived fuels in the Asia-pacific region.

DOI: 10.4324/9781003281818-4

4.2 The Role of Hydrogen in Decarbonising India's Energy Mix

India is committed to a low-carbon pathway for achieving accelerated and sustainable economic growth. This necessitates the share of renewable sources in electricity and heat generation, manufacturing and construction, and the transport sector with progressively less reliance on fossil fuels. Accordingly, one central task to fulfilling India's revised commitments under the 2015 Paris Climate Agreement is to reduce greenhouse gas (GHG) emissions intensity by 45 per cent below 2005 levels and achieve 50 per cent of installed electric power capacity from non-fossil sources by 2030.

In line with this ambition, the Government of India has established several policies for an aggressive roll-out of renewable energy (RE) with at least 500 GW installed capacity by 2030 for electric power generation, the electrification of the transport sector, and an industrial energy efficiency programme (Perform, Achieve, and Trade scheme), among other mitigation action. These policies target the sectors responsible for most emissions in the Indian economy.

In a business-as-usual scenario, coal is expected to drive the majority of future emissions growth in the power sector (Chaturvedi et al., 2018). The electrification of the transport sector shifts the GHG emissions source from vehicles to the power sector. This might further increase coal and other fossil demand because substituting current baseload electricity generation with renewables would require efforts to reduce the intermittency (both daily and seasonal) and associated interconnection costs. In the industrial sector, coal is already the cheapest fuel and abundantly available in India (Gupta et al., 2019). Substituting coal would require a low-carbon alternative to cater to the high-temperature heat requirement, particularly in the hard-to-abate sectors.

Against this backdrop, hydrogen has the potential to play a critical role in the economy's deep decarbonisation trajectory. Hydrogen as an energy carrier would allow RE to be distributed across regions, seasons, and sectors. Hydrogen produced from renewable sources of energy would serve multiple usage streams as fuel or via reconversion into electricity. Hydrogen fuelled automobiles can complement battery electric vehicles (BEVs) to achieve transport sector-wide decarbonisation objectives, while improving air quality. In industry, hydrogen can help curb emissions from processes that are hard to electrify and depend upon input materials or feedstock derived from fossil fuels, such as the production of fertilisers and other chemicals, petrochemicals, iron and steel. Existing natural gas infrastructure could be adapted to carry hydrogen and provide a means to reduce natural gas consumption and avoid additional investments to build new hydrogen pipelines.

4.3 Shades of Hydrogen

Hydrogen can be produced using various raw feedstock and sources of energy. Hydrogen has been given multiple colours to account for these different production pathways as detailed in Table 4.1.

Table 4.1 Shades of hydrogen.

Shades of hydrogen	*Source*	*Technology/ Process*	*Cost in/kg (year estimate)*	*Carbon footprint (kg CO_2/kg of hydrogen)*	*Reference*
Black/ Brown	Coal	Coal gasification (CG)	0.9–1.5	15–26	Parkinson et al., 2019
Grey	Natural gas	Steam methane reforming (SMR)	1.7–2.3	9.3–15	Parkinson et al., 2019; Biswas et al., 2020
Blue	Fossil fuel + carbon capture and sequestration (CCS)	CG/SMR + CCS	1.3–3.6 (CG) 2.7–3.3 (SMR)	2.1–6.2 (CG) 2.97–5.61 (SMR)	Parkinson et al., 2019; Biswas et al., 2020
Turquoise	Natural gas	Pyrolysis	0.9–3.5	4.2–6.1	Parkinson et al., 2019
Pink	Nuclear	Electrolysis	1.95–6.2	4.2–7	Parkinson et al., 2019
Green	Renewable power	Electrolysis	3.5–5.8	2	Biswas et al., 2020; Yadav & Banerjee, 2020

At present, the global consumption of hydrogen of all colours stands at roughly 88 million tonnes. With declining renewable tariffs, India is expected to become a producer of the most affordable green hydrogen globally (IEA, 2019).

4.4 Levers of a Green Hydrogen Transition

Hydrogen has gone through several phases in the global energy landscape. The first modern-day trigger point for hydrogen was the 1970s oil shortages and price shocks when hydrogen technology collaborations were initiated and expectations were raised. In the 2000s, these heightened expectations culminated in a narrow focus on hydrogen fuel cell vehicles. However, with oil prices rapidly falling and the financial crisis of the second half of the decade, the economic case for hydrogen weakened. Since the 2010s, hydrogen is gradually regaining momentum with a renewed focus on and urgency of the climate change challenge (Hall et al., 2020). The current phase is driven by a shifting focus to decarbonise the hard-to-abate sectors like heavy industries and transportation. Over the past two years, there has been a rapid shift in the global narrative, with approximately 32 countries and the European Union announcing their national hydrogen strategies, and the count is still increasing (Ghosh & Chhabra, 2021). However, a successful transition would require several actors supporting a wide-scale deployment and

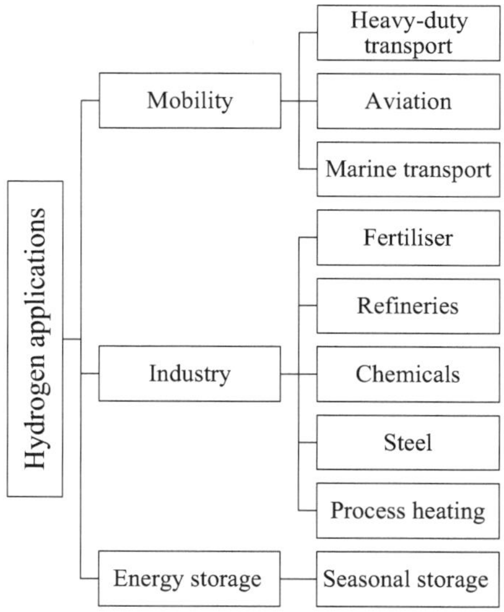

Figure 4.1 End-use applications of green hydrogen.

Source: Council on Energy, Environment and Water (CEEW) analysis

commercialisation of the technology. The levers of the transition to a hydrogen economy can be broadly classified as demand side, supply side, research and development (R&D), and other intermediary linkages, which the following sections will elaborate on further.

4.4.1 Demand Side: Early Market Creation

On the demand side, anticipating favourable government policies and the business attractiveness of the hydrogen economy, several leading companies have announced their business plans. India's largest conglomerate, Reliance Industries, for example, announced their US$10 billion investment towards building renewable and hydrogen technologies. In addition, there have been significant investment plans from central public sector undertakings and foreign direct investment inflows.

4.4.1.1 India's Green Hydrogen Opportunity

India significantly depends on imported fossil fuels to meet its energy needs. During 2019–2020, India imported US$112 billion of crude oil, US$21 billion of coal, and US$9 billion of liquefied natural gas (LNG) (MoPNG, 2020; CCO, 2020). Green hydrogen, by contrast, offers the long-term opportunity to decarbonise the economy and help achieve self-reliance and energy independence, at least in a few sectors. Figure 4.1 shows various end-use applications of green hydrogen.

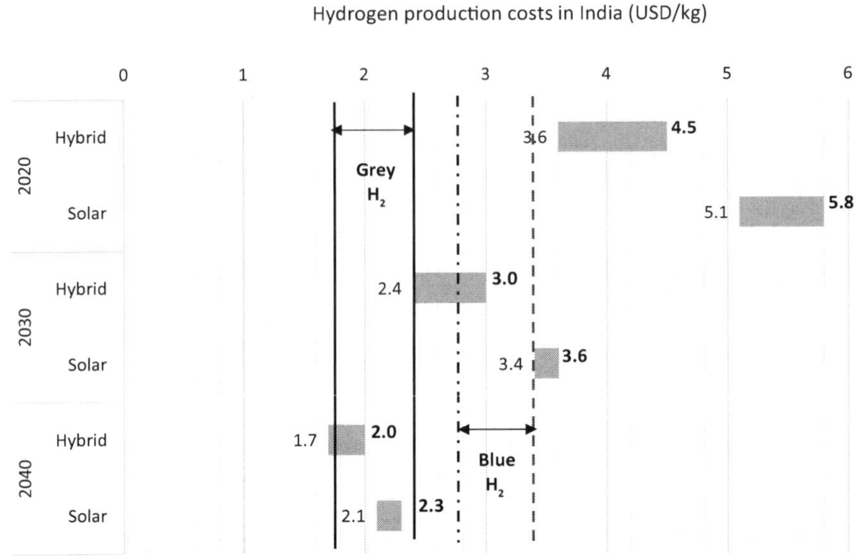

Figure 4.2 Green hydrogen production costs in India.

Source: Biswas et al. (2020)

4.4.1.2 Techno-Economics of Green Hydrogen Production in India

In 2020, the cost of producing green hydrogen from solar and wind energy sources is expected to be 3.5–5.8 USD/kg of hydrogen as depicted in Figure 4.2 (Biswas et al., 2020; Hall et al., 2020). The cost is expected to reduce to 2.4–3.6 USD/kg by 2030. At these prices, green hydrogen becomes cheaper than blue hydrogen and is price competitive with grey hydrogen produced from imported natural gas (LNG). Further, by 2040, green hydrogen is expected to break even with grey hydrogen obtained from natural gas. With a favourable policy ecosystem, India can become a dominant player in the global green hydrogen economy, catering to both domestic and export markets, particularly in the Asian subcontinent.

4.4.1.3 Existing Markets

Green hydrogen is a promising option for end-use applications like seasonal storage of electricity, heavy-duty transport, and hard-to-abate industrial sectors. However, before these over-the-horizon technologies commercialise, it is imperative to identify existing market opportunities for green hydrogen. Today, hydrogen is already used in industries to produce ammonia which is subsequently processed to obtain fertiliser, refine crude oil, and produce methanol. The Chlor-Alkali industry yields hydrogen as a by-product which is used as a fuel in flake plants and boilers or is sold in the market as compressed hydrogen (BEE, 2015).

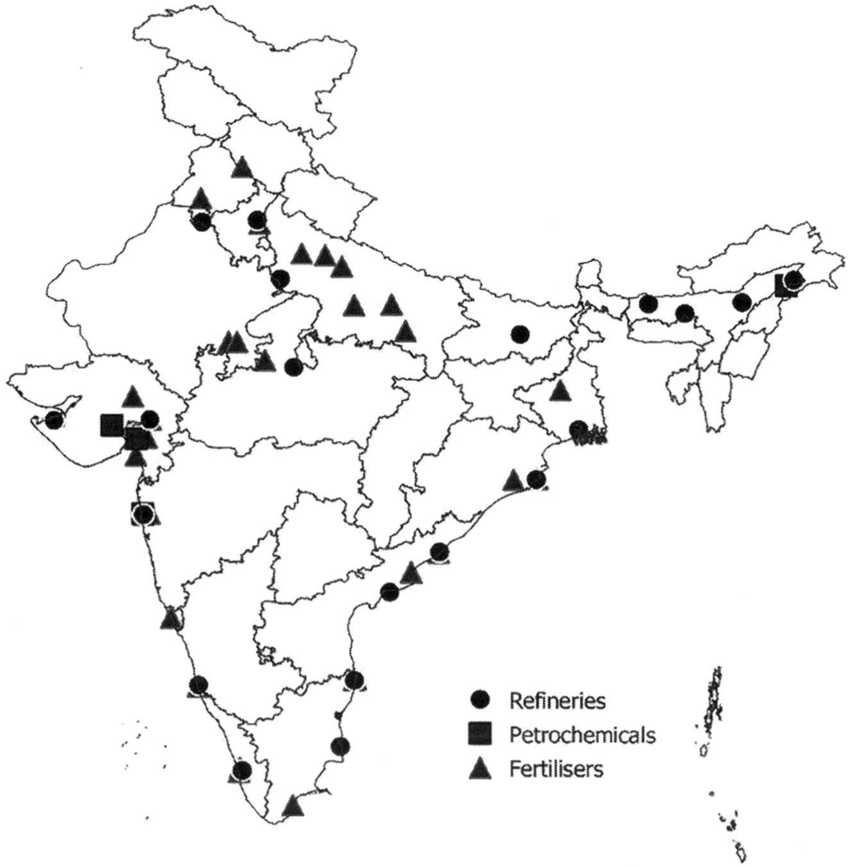

Figure 4.3 Green hydrogen consumption potential in the industry sector.

Source: Authors' own

In addition to the existing demand, there is also a latent demand for hydrogen in the chemicals and steel sector. India imported 2.27 million tonnes of methanol and 11 million tonnes of urea during 2019–2020 (DoC, 2020). Through hydrogen, these commodities can also be produced in India by utilising domestic resources. Similarly, hydrogen can replace 30 per cent of the gas consumption for making sponge iron in existing shaft furnaces without any modification to existing plants (Ripke & Kopfle, 2017). This latent demand would translate into additional hydrogen demand of 1.9 million tonne per annum (MTPA). Figure 4.3 illustrates the existing hydrogen demand across various end-use sectors in India.

The pre-2020 grey hydrogen demand in India of 5.6 million tonnes is roughly equivalent to a market size of US$11 billion.[1] With the inclusion of latent hydrogen demand, total hydrogen demand could grow more than fourfold between 2020 and 2050, amounting to around 29 million tonnes by 2050 (Niti Aayog & RMI, 2022).

4.4.1.4 Emerging Opportunities of Green Hydrogen

The hydrogen market in India is expected to increase from roughly 5.6 million tonnes in 2020 to around 11 million tonnes by mid-century. This growth primarily considers the existing consumers of hydrogen – fertilisers, refineries, and methanol. However, with the right policy push, hydrogen has the potential to substitute the incumbent fossil fuels. In 2050, the steel industry and import substitution of fertiliser and methanol can further increase the total hydrogen demand from 11 million tonnes to 20 million tonnes. Additional hydrogen demand can be unlocked in other end-use applications from transportation to RE storage.

A. RE Grid Integration and Storage

Sustained efforts have already seen the installation of over 100 GW of renewable capacity, and the share of non-hydro RE in the electricity mix rose to about 10 per cent in the year 2018–2019 (MoSPI, 2019). As the penetration of renewables in India's electricity mix increases, there is an impending need for technological solutions balancing the intermittent power production from renewables while maintaining the stability and safety of the electricity grid and avoiding curtailment. Energy storage technologies can assist the network operators in maintaining this demand–supply balance in the grid by providing ancillary services like ramping (up and down) and frequency regulation. For long-duration storage running for several hours, converting excess available energy into hydrogen and utilising it for grid support and other applications is perceived as a suitable option. The Energy and Resources Institute's analysis indicates that hydrogen storage can become competitive with battery storage technologies for long-term (weekly/monthly/seasonal) storage applications by 2030 (Hall et al., 2020). However, the potential long-term storage requirement will be highly dependent on the extent of solar and wind energy deployment.

B. Transportation

The Department of Heavy Industries and NITI Aayog have been coordinating efforts to push electric mobility in the country with an aim to reduce dependence on imported fossil fuels, curb local pollution, and reduce carbon emissions from the transport sector. Fuel cell electric vehicles (FCEVs) offer a value proposition similar to BEVs in terms of reduced emissions while also offering significant advantages in terms of longer range and faster refuelling time. FCEVs run on hydrogen fuel to produce onboard electrical power for propulsion with zero tailpipe emissions. If hydrogen is produced from an RE source, FCEVs would have a negligible carbon footprint. However, in the present assessment, FCEVs fuelled by green hydrogen are not cost-competitive with BEVs or conventional fossil fuel-based vehicles across most segments. The International Energy Agency (IEA) estimates (2019) indicate that FCEVs can break even with BEV for vehicles having a driving range of 400 km or for interstate long-distance travel. Studies have also indicated that for heavy-duty vehicles, such as commercial vehicles with longer trip range, route flexibility, and negligible detriment to payload capacity, FCEVs could be cost-competitive in the very near term. The IEA estimates further that with higher utilisation of infrastructure assets the total cost of ownership for a fuel cell truck breaks even with its BEV equivalent by 2030 (IEA, 2019). However, in the long term, a continued decline in battery costs and improvement in energy density will increase the competitiveness of battery-electric heavy-duty vehicles vis-à-vis their fuel cell counterpart.

Green hydrogen can also reduce emissions in the aviation industry. Currently, aviation contributes 2.4 per cent of global GHG emissions (EESI, 2019). Green hydrogen obtained from solar and wind energy sources can be subsequently used to produce sustainable aviation fuel (SAF) in a Fischer-Tropsch process. However, the cost of SAF is prohibitively high for the industry to adopt directly. At a green hydrogen price of 3.2 USD/kg, the cost of SAF is 2,000 USD/tonne (WEF, 2020). By contrast, the cost of fossil-based aviation fuels today is 400–700 USD/tonne. The American Society for Testing Materials allows up to 50 per cent blending of SAF in the aviation industry. This suggests a pathway for the incremental decarbonisation of the industry by blending SAF in existing fossil-based jet fuel.

Hydrogen will also play a critical role in decarbonising the shipping industry. Achieving the International Maritime Organisation's goal of at least a 50 per cent reduction in total GHG emissions from the sector would require a shift away from conventional fossil fuels. Among the various low-carbon alternatives, ammonia derived from green hydrogen will have the lowest total cost of ownership (IEA, 2019). However, the uptake of green ammonia as a shipping fuel would need engine modification and bunkering facilities at ports.

C. Industry

The industrial sector is the single largest consumer of fossil sources of energy in India and, as a result, a significant contributor to GHG emissions. The heat demand in manufacturing processes can be decarbonised by using low-carbon fuels, like natural gas, or renewable fuels, like biomass and green hydrogen. Carbon capture and storage is also a likely pathway to reduce emissions from the sector. There are significant technical challenges and economic viability constraints for adopting these technologies in the manufacturing sector. However, the industry plays a critical role by providing the volumes to achieve economies of scale necessary to drive down costs of several types of equipment in the hydrogen supply chain.

Hydrogen can potentially replace coal and coke in iron and steel production, thereby mitigating the single largest emission's source in the industry. Hydrogen would also help reduce the import bill of the country by reducing coking coal imports by the steel industry. The chemical industries are already-existing consumers of hydrogen used for the manufacturing of ammonia, methanol, hydrogenation of oil products, and the like. These industries are currently producing hydrogen by reforming natural gas. Substituting natural gas with hydrogen produced from water electrolysis would shave off a significant subsidy burden of the government. Various studies have indicated that with a favourable decline in electrolyser CAPEX and renewable electricity prices, green hydrogen can start replacing fossil fuels, particularly in the ammonia and steel manufacturing process (Nayak-Luke & Bañares-Alcántara, 2020; Yadav et al., 2021).

A 100 per cent green hydrogen-based steelmaking will become competitive with the conventional blast furnace process only in 2040. Green hydrogen steel can become competitive with the existing coal-based processes in a blast furnace if the cost of green hydrogen can be brought down to 1.3–2.2 USD/kg (Yadav et al., 2021). However, such reductions require aggressive reduction of the price for electrolysers and renewable electricity. An alternate approach to incentivise the uptake of green hydrogen in the steel industry is by incremental blending with grey hydrogen. With a 9 per cent blend of green with grey hydrogen, the steel production costs are already price-competitive with the conventional process (Yadav et al., 2021). Further, as electrolyser and renewable electricity prices decrease, the share of green hydrogen can be gradually increased to 60 per cent by 2030 and 100 per cent by 2040.

Several studies have already indicated that green ammonia has the potential to become affordable within the next decade (Hall et al., 2020; Nayak-Luke & Bañares-Alcántara, 2020). However, the cost competitiveness of green ammonia highly depends on access to favourable renewable profiles. This is particularly true for locations in the western part of the country that will be able to produce green ammonia at less than US$400 per tonne, thus breaking even with the conventional production process that works with subsidised natural gas prices of US$ 6.5 per MMBtu.

The majority of ammonia produced in the country today is further converted to urea. In the conventional process, natural gas is reformed to produce hydrogen and carbon dioxide (CO_2). The hydrogen is utilised to produce ammonia, which is then further reacted with CO_2 to produce urea. Most of the ammonia plants are directly coupled with the urea production unit. Substituting the conventional process with green ammonia would require an "external" CO_2 source. Hence, production costs would increase depending on the cost of capturing, transporting, and perhaps storing CO_2. This would most probably result in increasing the subsidy burden of the exchequer.

However, one possible transition pathway can be to start using green ammonia to produce other direct ammonia-based fertilisers such as calcium ammonium nitrate and diammonium phosphates. In 2020, India imported nearly 65 per cent of its complex fertiliser (diammonium phosphate).

Increasing the domestic availability of complex fertilisers would reduce the over-utilisation of high subsidised urea in farmlands to achieve the ideal nutrient ratios. In Punjab, for instance, the nitrogen:phosphorus:potassium ratio stands at 26:6:1, compared to an ideal ratio of 4:2:1.

Hydrogen can also be used to produce methanol. Unlike the fertiliser industries, methanol production units do not have access to subsidised natural gas and are exposed to international LNG import price volatility. More than 90 per cent of the country's current methanol demand is imported. In an attempt to reduce import dependency, the Government of India is promoting the use of CG technology. However, with an aggressive reduction of electrolyser costs of 300 USD/kW and renewable electricity prices (solar and wind electricity prices of US$27 and US$41 per MWh, respectively), the cost of green hydrogen-based methanol at 400/tonne becomes competitive with CG (Biswas et al., 2019; Hall et al., 2020).

4.4.1.5 Early Market Penetration With Blending Mandates

Putting the latent hydrogen demand onto a green basis would require policy interventions favouring the creation of a flexible market. As pointed out earlier, the high initial cost of 100 per cent green hydrogen can be significantly brought down by blending with grey hydrogen in already-existing processes. By contrast, 100 per cent dispatchable green hydrogen production requires significant investments in RE storage in the form of both hydrogen and battery in order to overcome the supply gaps created by RE intermittencies. A blended hydrogen production minimises these costs by producing green hydrogen whenever renewable electricity is available and flexibly switching back to grey hydrogen in its absence. As the electrolyser and storage costs decline, incremental blending targets can be achieved with a marginal increase in overall production costs. To guarantee demand, the Ministry of New and Renewable Energy (MNRE) is already considering putting in place green hydrogen consumption obligation mandates. Such a mandate would support the early market development and help achieve the economies of scale necessary for cost reduction in green hydrogen production. We estimate that, by 2030, the policy can support nearly 2.2 million tonnes of green hydrogen consumption across various end-use applications as shown in Table 4.2.

This potential demand can support nearly 60 GW of solar PV capacity and another 40 GW of electrolyser capacity. The market would also attract investments worth US$56 billion in renewables and electrolysers and an additional US$9 billion in downstream activities, such as the establishment of hydrogen refuelling stations (HRSs) and industrial process retrofits. The following assumptions were made to arrive at these estimates:

- For the fertiliser industry, we assume that all ammonia production facilities will use a 20 per cent blend of green hydrogen in their existing process at no significant greenfield investment.

Table 4.2 Green hydrogen demand, infrastructure requirement, and investment need by 2030.

	Green hydrogen demand	*Size (GW)*		*Hydrogen production investment*	*Downstream investment*
Sector:	MTPA	Electrolyser	Photovoltaic (PV)	Billion (US$)	Billion (US$)
Fertiliser	0.85	15.20	22.79	21.64	0.00
Refinery	0.80	14.38	21.56	20.47	0.00
Methanol	0.03	0.57	0.85	0.81	0.00
Steel	0.45	8.04	12.06	11.45	0.00
Transport	0.06	1.09	1.64	1.55	8.96
Total	2.20	39.28	58.89	55.92	8.96

Source: Authors' analysis

- Refineries produce hydrogen either by using natural gas or by cracking naphtha and other oil products. We assume that, by 2030, all refineries will achieve a green hydrogen blending target of 25 per cent. A similar target has also been considered for methanol production.
- India meets the targets identified in the national steel policy for the year 2030. The share of gas-based technologies in 2030 is 34 MTPA. It is assumed that hydrogen-based steelmaking will reach parity with the gas-based process. The gas-based shaft furnace can absorb up to 30 per cent of the hydrogen – on an energy basis – without modifying the plant. Therefore, in 2030, it is assumed that 50 per cent of the gas-based plants will start blending hydrogen with gas.
- For the transport sector, the projections for the number of vehicles across various modes are obtained based on CEEW estimates (Soman et al., 2020). The penetration of hydrogen vehicles in the future is taken as the world average for future scenarios (IEA, 2018). Within each vehicle segment, the relative share of FCEVs is based on the estimates for developing markets like China (Samsun et al., 2020). The hydrogen demand is estimated by considering the average distance travelled by each vehicle type in a year and the respective specific fuel consumption.

While the uptake of green hydrogen has benefits in a reduction in GHG emission footprint and import dependency, there are also challenges regarding huge capital investments. An uptake of 2.2 MTPA of green hydrogen could reduce India's natural gas import by 7.5 MTPA and mitigate 19.37 MTPA of CO_2 emissions (IEA, 2017). This would create a demand for 60 GW of solar power and 40 GW electrolysers, resulting in investment requirements of about US$65 billion.

4.4.2 Supply-Side Ecosystem Development

Foundational technologies, such as green hydrogen, need a supportive ecosystem to compete with incumbent fossil fuels and feedstock. The hydrogen value chain consists of various stages – from production, storage, transportation to end-use. Each stage is dependent on different technologies. Ever since the year 2006 when India's first National Hydrogen Energy Road Map was prepared, MNRE has been supporting technology development across the hydrogen value chain. Between 2010 and 2020, financial support of around US$13 million was extended to various research institutions, academia, and industries. These research projects have resulted in the successful development of fuel cell stacks, hydrogen-fuelled vehicles, and HRSs (DST, 2020). In addition, several competitive technologies have been developed for hydrogen production from biomass and for storage in metal hydrides and carbon materials. Alongside the MNRE's initiatives, major institutions like the Defence Research and Development Organisation, the Bhabha Atomic Research Centre (BARC) and the Indian Space Research Organisation have been developing cutting-edge hydrogen and fuel cell technologies. The Department of Science and Technology has also been supporting basic R&D in hydrogen and fuel cells.

4.4.2.1 Electrolyser Manufacturing

At present, there are two major commercially available electrolyser technologies – alkaline electrolyser (AE) and proton exchange membrane (PEM) electrolysers. Simultaneously, alternatives such as anion exchange membrane, solid oxide electrolysers, and electrochemical, thermally activated chemical processes are still in various stages of development. These electrolysers can be directly coupled to RE systems for producing green hydrogen. The key technical and cost parameters for the alkaline and PEM electrolysers are indicated in Table 4.3.

Table 4.3 Technical and cost outlook for AE and PEM electrolysers.

Parameter	*AE*			*PEM electrolyser*		
Year	2020	2030	2040–2050	2020	2030	2040–2050
Electricity consumption (kWh/kg hydrogen)	48–53	47–51	42–48	50–60	49–53	45–50
Life (hours)	60–90	90–100	100–150	50–80	60–90	100–150
System cost (/kW)	500–1.000	400–850	200–700	700–1,400	650–1,500	200–900

Source: IEA (2019); IRENA (2020)

While various measures are put in place to create demand for green hydrogen in industry and beyond, the availability of low-cost, high-performance electrolysers is crucial to reduce the production cost of green hydrogen. Initial estimates suggest that the potential demand for green hydrogen would require gigawatt-scale domestic manufacturing capabilities for electrolysers (Mayyas et al., 2018). Implementing the green hydrogen consumption obligations would require electrolyser capacities of 40 GW by 2030. This capacity is sufficient to provide the necessary economies of scale for electrolyser manufacturing to slash their existing costs by more than 80 per cent. Several major manufacturers have already announced aggressive cost reduction targets (IRENA, 2020). India has the potential to manufacture low-cost electrolysers by leveraging the existing manufacturing base of power electronics, compressors, and other balance of plant components achieving economies of scale from expected domestic demand and potential export markets.

The BARC, for example, has developed high-performance alkaline water electrolysers based on advanced electrolytic modules using nickel electrodes. A portable electrolyser of 1.5 Nm^3/h hydrogen-production capacity and large units of capacities between 10 and 30 Nm^3/h hydrogen production have also been developed. BARC is further working on the development and demonstration of PEM water electrolysers of 1.0 and 5.0 Nm^3/h capacity with energy consumption of about 5.75 kWh/Nm^3 of hydrogen at 5–10 bar pressure. The Centre of Fuel Cell Technology, Chennai, has developed and demonstrated a 1.0 Nm^3/h hydrogen-production-capacity PEM electrolyser using water and methanol as feedstock with a much lower energy consumption of 1.40 kWh/Nm^3. The Centre also demonstrated the world's first-time use of carbon-based materials in its construction. A PEM electrolyser system of 30 Nm^3/h capacity has been installed at the Indian Oil's R&D Centre, Faridabad, and an AE of 10 Nm^3/h capacity powered by a 120 kW solar PV plant is installed at the National Institute of Solar Energy (NISE) in Gurgaon (DST, 2020).

4.4.2.2 Hydrogen Storage

Hydrogen can be stored in gaseous form, liquid form by direct liquefaction, using liquid organic hydrogen carriers (LOHCs) or by processing hydrogen to produce ammonia and methanol. In gaseous form, hydrogen can be stored in salt/rock caverns, depleted gas fields, or pressurised containers. The conversion of hydrogen into a liquid state is preferred for long-distance transport. However, the conversion also has an energy penalty and monetary implication. According to the IEA (2019), the typical energy penalty, which is represented in the percentage of the lower heating value of hydrogen, ranges between 25 and 35 per cent for liquefaction, 7 and 18 per cent for ammonia, and 35 and 40 per cent for LOHCs. Compared to gaseous state and liquid state storage of hydrogen, the solid-state storage in metal hydrides is technologically less ready. Table 4.4 summarises the comparison of various routes for storing hydrogen.

Table 4.4 Hydrogen storage costs across various technologies.

	Gaseous state				*Liquid state*			*Solid state*
	Salt caverns	Depleted gas fields	Rock caverns	Pressurised containers	Liquid hydrogen	Ammonia	LOHCs	Metal hydrides
Usage volumes	Large	Large	Medium	Small	Small-Medium	Large	Large	Small
Cycling times	Weeks–months	Seasonal	Weeks–months	Daily	Days–weeks	Weeks–months	Weeks–months	Days–weeks
Benchmark LCOS* (/kg)	0.23	1.90	0.71	0.19	4.57	2.83	4.50	–
Possible future LCOS (/kg)	0.11	1.07	0.23	0.17	0.95	0.87	1.86	–
Geographic availability	Limited	Limited	Limited	Not Limited	Not Limited	Not Limited	Not Limited	Not Limited

Note: * Levelised cost of storage (LCOS).

Source: BNEF (2020)

The Indian Institute of Technology (IIT) Madras and Banaras Hindu University have identified a wide range of materials such as carbon and hydrides that facilitate efficient hydrogen storage and are cyclable, stable, and capable of being utilised in mobile applications. IIT Madras primarily focuses on storage in carbon materials, while Banaras Hindu University focuses on storage in hydrides. The technology has been installed in two-wheeler, three-wheeler, and a car for performance testing and for demonstration of proof of concept.

4.4.2.3 Transportation and Distribution Infrastructure

The supply and demand nodes of green hydrogen are located in different geographies. Therefore, hydrogen has to be distributed from production centres to demand points. The various options are: (1) hydrogen can be distributed in either gaseous or liquid form in tankers; (2) hydrogen-ready pipelines can move gaseous hydrogen over long distances at affordable prices; (3) hydrogen can be transported through chemicals like LOHCs; and (4) hydrogen can be further processed to portable ammonia or methanol.

(1) Tankers

Hydrogen transport in tankers and trucks might be a viable option for short distance and low-volume hydrogen transport. In trucks, hydrogen can be moved in gaseous and liquid forms. In gaseous form, hydrogen is moved in pressurised tube trailers by compressing it to 200–500 bar. While the transport cost of liquid hydrogen is significantly lower than moving it in a gaseous state, the liquefaction step consumes a large amount of energy. Also, liquid hydrogen transportation can result in boil-off losses if sufficient temperature inside the containers is not maintained (Hall et al., 2020).

(2) Pipelines

Hydrogen transport in pipelines offers an efficient way of moving hydrogen in gaseous form. However, the challenge with the hydrogen pipeline is the high initial investment. It is observed that the tariff of moving hydrogen in pipelines depends on the distance and the pipeline capacity. A challenge with hydrogen pipelines is that it needs huge capital investment to lay the pipeline. The transmission tariff is also dependent on the utilisation factor of the pipeline. If the utilisation factor is low, the transmission tariff is high. Therefore, a prerequisite for the hydrogen pipelines would be a guaranteed offtake of renewable hydrogen that can justify the high capital investment.

(3) Derivative Chemicals

Hydrogen can also be moved in the form of LOHCs and ammonia to various demand nodes via trucks. The cost of transporting hydrogen in these energy

Table 4.5 Transport tariffs across different modes and distances.

Distance (km)	*Pipeline capacity*		*Trucks (gaseous hydrogen)*	*Trucks (liquid hydrogen)*	*Trucks (LOHCs)*	*Trucks (ammonia)*
	100 TPD	*500 TPD*				
50	0.12	0.023	0.48	0.13	0.18	0.07
100	0.2	0.06	0.61	0.16	0.26	0.12
500	0.88	0.293	1.84	0.39	0.82	0.43

Source: IEA (2019)

carriers is lower than that of gaseous hydrogen. However, the overall economics should consider the conversion and reconversion losses with liquid hydrogen carriers. It is also important for India to evaluate the economics of moving green hydrogen via railways, especially for long-distance transport. Table 4.5 indicates the tariffs of moving hydrogen in per-kilogram of hydrogen for two different pipeline capacities – 100 and 500 tonnes per day (TPD) – and using trucks over a transport distance of 50 to 500 km (IEA, 2019).

(4) Refuelling Stations

The availability of HRSs has a very strong influence on the adoption of fuel cell vehicles. The actual value of refuelling stations will depend on multiple factors like population density, purchasing power, and vehicle penetration. In general, it is expected that 100 HRS units can support 100–400 FCEVs per station, and 1,000 HRSs can support 1,000–2,000 vehicles per station (ICCT, 2017).

Figure 4.4 shows the break-up of the current delivered cost of hydrogen at a refuelling station. It can be seen that the refuelling station itself forms the bulk of the levelised cost of hydrogen delivered to a vehicle. The hydrogen compressor and the storage unit are the most expensive components of the HRS. Hence, in order to incentivise the uptake of FCEVs, a reduction in the cost per HRS unit is critical. It is expected that this reduction will happen with an economy of scale and larger capacity of the refuelling station. Studies indicate that the levelised cost of refuelling stations can reduce from 6 USD/kg to less than 4 USD/kg if the refuelling station size increases from 200 kg/day to 1,000 kg/day. With higher deployment volumes, this cost can reduce to less than 2 USD/kg (Reddi et al., 2017).

Currently, demonstration and pilot-testing of HRSs are carried out by the NISE and the R&D Centre at Indian Oil Corporation (IOC). The HRS at the IOC centre is currently supporting the pilot-testing of hydrogen-fuelled buses developed by TATA Motors, Mahindra & Mahindra, and IIT Delhi.

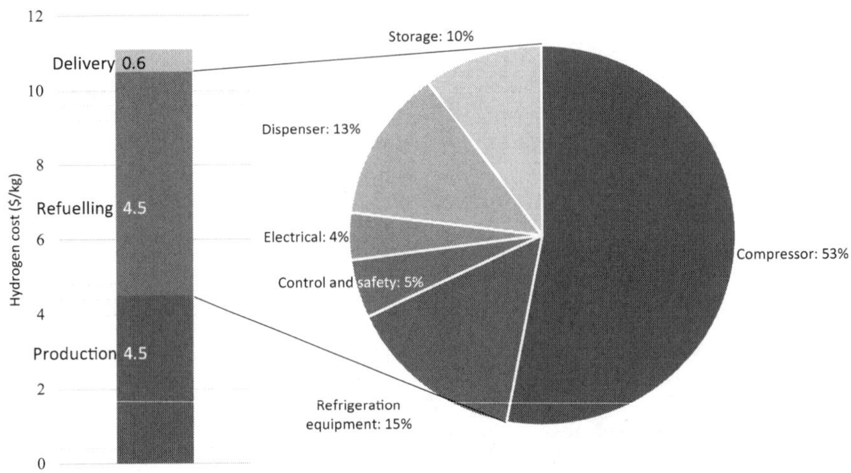

Figure 4.4 Break-up of the total delivered cost of hydrogen at refuelling station (200–300 kg/day capacity).

Source: Reddi et al. (2017)

4.4.3 *Technology and Knowledge – R&D*

The existing R&D ecosystem is frail and non-conducive towards pioneering research in innovative horizon technologies. In 2018, the share of R&D investments in national gross domestic product stood at merely 0.65 per cent compared to a global average of 2.3 per cent (World Bank, 2020). Horizon technologies like green hydrogen have significant emissions reduction potential but simultaneously have high investment costs and associated risks of potential technology failures. Hence, in order to incentivise technology development, the costs and the associated risks are required to be shared. A public–private partnership (PPP) approach can facilitate collaboration between industry, government, and academia, often extending beyond the national boundaries. Globally, several of the hydrogen projects have considered a PPP approach. Denmark, for example, was able to raise EUR 28 billion through a PPP financing model to develop an Energy Island in the middle of the North Sea. The island would support a 3-GW electrolyser capacity. A further expansion to 10 GW is anticipated (Durakovic, 2020).

4.4.3.1 Dedicated R&D Fund

The scale of innovations required for developing a portfolio of clean, efficient, and market-ready solutions, which enables exploiting the properties of hydrogen as energy carriers, would require much more investments in R&D. Our estimates indicate that, by 2030, India would need investments of US$55 billion to commercialise 2.2 million tonnes of green hydrogen.

Thus, there is a need for a dedicated R&D fund, which should have 50 per cent contributions from industry and 50 per cent contribution from the government. In addition to supporting technology demonstrations, the fund can also ensure the regular award of a grant to centres of excellence focusing on medium- and long-term research supporting potential breakthroughs in the sector. The availability of R&D funds would also give a lot of autonomy, promoting proposals and ideas that focus on more short-term, applied research in the field. Such grants via the R&D programme can provide the necessary seed fund required for the development of start-ups and small- and medium-sized enterprises in the ecosystem.

4.4.3.2 Centres of Excellence

Establishing Centres of Excellence on hydrogen should be envisioned under the R&D facilitation strategy. The centres will play an important role in developing the expertise and capability required for scaling hydrogen. It will be responsible for carrying out research for long-term breakthroughs and research on applied applications of the technology in the short and medium terms. It will also enable knowledge sharing with international universities and centres of excellence, further boosting capacities, capabilities, and expertise. In addition, incubation centres could nurture start-ups that focus on improving the economics of hydrogen adoption and production.

4.4.4 Intermediary Linkages

Intermediary linkages refer to the interconnection between the various hydrogen ecosystem components. The cross-sectoral applications of hydrogen require establishing a robust governance framework for its deployment and scaling. The framework should be able to support a wide range of activities such as facilitating a policy environment, standards, and regulations, demonstrating pilots and developing necessary infrastructure. While the National Hydrogen Mission aims to establish such a framework, its implementation would require a significant heavy lifting and alignment of vision across the various line ministries.

There is also a need for global leadership for concerted effort in building a global green hydrogen economy. Ghosh and Chhabra (2021) have analysed 24 countries and regions and found that they rarely involve a developing country for bilateral or plurilateral partnerships on various aspects of green hydrogen technology and market development. Ironically, developing economies are expected to drive the majority of future energy demand. Establishing a Global Green Hydrogen Alliance would have to ensure equity in terms of access to technology while simultaneously sharing the burden of failures in technology development or second- and third-order risks such as governance failures and trade or intellectual property disputes associated with deployment at scale.

4.4.5 Critical Actors for Successful Implementation

MNRE has been entrusted with drafting and implementation of India's Hydrogen Mission. The hydrogen economy inherently is an extremely complex ecosystem. While hydrogen has cross-sectoral applications of hydrogen, its technical standards can have a wide variation. For example, applications in transport require extremely pure hydrogen (>99.99 per cent by volume) when compared to applications within the steel industry. Similarly, deploying hydrogen infrastructure that extends well beyond the existing captive consumption within industry boundaries require careful evaluation of regulations and anticipation of safety concerns across its entire value chain. Establishing a robust mission governance system becomes extremely important in order to establish close coordination between the various line ministries, departments, state governments, industry, regulatory and scientific institutions, and other relevant stakeholders. The governance structure can be constituted by the mission secretariat (MNRE), a standing committee (constituted with line ministries, NITI Aayog, and regulatory agencies), and an expert group (constituted with research organisations, academic institutes, and civil society).

4.5 Conclusion and the Way Forward

From the above discussion, it is clear that there already exists a large hydrogen market in India. However, green hydrogen is still not competitive with existing industrial processes, and the price barrier remains the central bottleneck for its wider adoption. A way forward would be to formulate a policy that incentivises the blending of green hydrogen within the existing end-use applications. This would have manifold benefits: industries will only pay a marginal premium for a switch to clean fuel, and the gradual scaling up of green hydrogen will reduce the component costs due to the learning rate effects. Such a smooth transition would unlock new market opportunities in transport and industrial sectors.

The next step would be to focus on upcoming opportunities in transport, steel and, blending hydrogen in gas pipelines. For transport applications, the focus should be on reducing the cost of HRSs by indigenising various components like compressors and power electronic units. Since creating the hydrogen infrastructure is a challenge for transport applications, India could benefit from blending green and grey hydrogen to reduce the initial price barrier in transportation. As the hydrogen economy scales, the share of green hydrogen can be incrementally increased over the years. Similarly, for hydrogen-based steelmaking, it is important that India prioritises investments in hydrogen-ready technologies like shaft furnaces that can eventually shift to green hydrogen. Hydrogen adoption in gas pipelines is challenging as it will need modifications in the pipelines and on the end-consumer side.

It is important that India takes the lead to become a first-mover on green hydrogen by leveraging its vast and cost-effective renewable portfolio. India should prioritise investments in research to indigenise the manufacturing of various components in the hydrogen value chain. A collaboration between industry-academia-research labs is needed to rapidly upscale India's Green Hydrogen Ecosystem.

Green hydrogen today is at a stage where solar energy was decades ago. A rapid policy push can encourage the industry and markets to quickly mobilise and break the current high-price and low-demand loop. But especially for India, it is an opportunity to become a US$5 trillion green economy by rapidly squaring the sustainability, economic growth, and employment objectives.

Acknowledgements

The authors acknowledge funding from Oak Foundation for this publication.

Note

1 We assume a hydrogen price at 1.7 USD/kg for fertiliser units (vs. natural gas at 6.3 USD/MMBtu) and 2.37 USD/kg for other industries (vs. natural gas at 11.5 USD/MMBtu).

References

BEE (2015). *Normalisation Document and Monitoring & Verification Guidelines – Chlor Alkali Sector*. Bureau of Energy Efficiency, Ministry of Power, Government of India, New Delhi. Retrieved from: https://beeindia.gov.in/sites/default/files/Chlor-Alkali-1-44.pdf (Last accessed: 15.02.2022)

Biswas, T., Ganesan, K. & Ghosh, A. (2019). *Sustainable Manufacturing for India's Low-Carbon Transition*. Council on Energy, Environment and Water, New Delhi. Retrieved from: www.ceew.in/publications/sustainable-manufacturing-indias-low-carbon-transition (Last accessed: 05.12.2021)

Biswas, T., Yadav, D. & Guhan, A.B. (2020). *A Green Hydrogen Economy for India – A Green Hydrogen Economy Production Cost*. Council on Energy, Environment and Water, New Delhi.

BNEF (2020). *Hydrogen Economy Outlook*. Bloomberg New Energy Finance. Retrieved from: https://data.bloomberglp.com/professional/sites/24/BNEF-Hydrogen-Economy-Outlook-Key-Messages-30-Mar-2020.pdf (Last accessed: 29.12.2021)

CCO (2020). *Coal Directory of India 2019–20*. Coal Controller's Organisation, Ministry of Coal, Government of India, Kolkata. Retrieved from: www.coalcontroller.gov.in/writereaddata/files/download/coaldirectory/CoalDirectory2019-20.pdf (Last accessed: 15.02.2022)

Chaturvedi, V., Koti, P.N. & Chordia, A.R. (2018). *Sustainable Development, Uncertainties, and India's Climate Policy – Pathways towards Nationally Determined Contribution and Mid-Century Strategy*. Council on Energy, Environment and Water (CEEW), New Delhi. Retrieved from: www.ceew.in/publications/sustainable-development-uncertainties-and-india%E2%80%99s-climate-policy (Last accessed: 15.02.2022)

DoC (2020). *Trade Statistics*. Department of Commerce, Ministry of Commerce and Industry, Government of India, New Delhi. Retrieved from: https://commerce.gov.in/trade-statistics/ (Last accessed: 10.12.2021)

DST (2020). *Compendium on Hydrogen and Fuel Cell 2018*. Department of Science & Technology, Government of India, New Delhi. Retrieved from: https://dst.gov.in/sites/default/files/2.%20Final%20Compendium%20-%20HFC%202018.pdf (Last accessed: 15.02.2022)

Durakovic, A. (2020). *Energy Island Could Power All Ships and Aircrafts in Denmark – Study*. Retrieved from: www.offshorewind.biz/2020/12/01/energy-island-could-power-all-ships-and-aircraft-in-denmark-study/ (Last accessed: 14.02.2022)

EESI (2019). *The Growth in Greenhouse Gas Emissions from Commercial Aviation. Fact Sheet*. Environmental and Energy Study Institute (EESI). 17.10.2019. Retrieved from: www.eesi.org/papers/view/fact-sheet-the-growth-in-greenhouse-gas-emissions-from-commercial-aviation (Last accessed: 12.11.2021)

Ghosh, A. & Chhabra, S. (2021). *Speed and Scale for Disruptive Climate Technologies – Case for a Global Green Hydrogen Alliance*. Council on Energy, Environment and Water, New Delhi. Retrieved from: www.ceew.in/publications/how-can-global-green-hydrogen-alliance-meet-international-clean-energy-needs (Last accessed: 23.09.2021)

Gupta, V., Biswas, T., Janakiraman, D., Jamal, A. & Ganesan, K. (2019). *Industrial Energy Use and Emissions Database*. Council on Energy, Environment and Water, New Delhi. Retrieved from: www.ceew.in/data/industrial-energy-emissions-dashboard (Last accessed: 15.02.2022)

Hall, W., Spencer, T., Renjith, G. & Dayal, S. (2020). *The Potential Role of Hydrogen in India: A Pathway for Scaling-Up Low Carbon Hydrogen across the Economy*. The Energy and Resources Institute (TERI), New Delhi. Retrieved from: www.teriin.org/sites/default/files/files/hydrogen-es.pdf (Last accessed: 05.11.2021)

ICCT (2017). *Developing Hydrogen Fueling Infrastructure for Fuel Cell Vehicles: A Status Update*. The International Council on Clean Transportation, Washington, DC.

IEA (2017). *Techno-Economic Evaluation of SMR Based Standalone (Merchant) Hydrogen Plant with CCS*. IEA Greenhouse Gas R&D Programme, Paris. Retrieved from: https://ieaghg.org/exco_docs/2017-02.pdf (Last accessed: 15.02.2022)

IEA (2018). *World Energy Outlook 2018*. International Energy Agency, Paris.

IEA (2019). *The Future of Hydrogen – Seizing Today's Opportunities*. International Energy Agency, Paris.

IRENA (2020). *Green Hydrogen Cost Reduction – Scaling up Electrolysers to Meet the 1.5°C Climate Goal*. International Renewable Energy Agency, Abu Dhabi.

Mayyas, A., Ruth, M., Pivovar, B., Bender, G. & Wipke, K. (2018). *Manufacturing Cost Analysis for Proton Exchange Membrane Water Electrolyzers*. National Renewable Energy Laboratory, Golden. Retrieved from: www.nrel.gov/docs/fy19osti/72740.pdf (Last accessed: 14.09.2021)

MoPNG (2020). *Indian Petroleum & Natural Gas Statistics 2019–20*. Ministry of Petroleum & Natural Gas, Government of India, New Delhi. Retrieved from: https://mopng.gov.in/files/TableManagements/IPNG-2019-20.pdf (Last accessed: 15.02.2022)

MoSPI (2019). *Energy Statistics 2019*. Ministry of Statistics and Programme Implementation, Government of India, New Delhi. Retrieved from: www.mospi.gov.in/documents/213904/301563//Energy%20Statistics%202019-finall1602098943619.pdf/b7e6e6b8-12da-7207-8f6a-b31ec54289a0 (Last accessed: 15.02.2022)

Nayak-Luke, R.M. & Bañares-Alcántara, R. (2020). Techno-economic viability of islanded green ammonia as a carbon-free energy vector and as a substitute for conventional production. *Energy and Environmental Science*, 2957–2966. https://doi.org/10.1039/D0EE01707H

Niti Aayog & RMI (2022). *Harnessing Green Hydrogen Opportunities for Deep Decarbonisation in India*. Retrieved from: https://www.niti.gov.in/sites/default/files/2022-06/Harnessing_Green_Hydrogen_V21_DIGITAL_29062022.pdf (Last accessed: 19.01.2024)

Parkinson, B., Balcombe, P., Speirs, J., Hawkes, A. & Hellgardt, K. (2019). Levelized cost of CO_2 mitigation from hydrogen production routes. *Energy & Environmental Science*, 19–40.

Reddi, K., Elgowainy, A., Rustagi, N. & Gupta, E. (2017). Impact of hydrogen refueling configurations and market parameters on the refueling cost of hydrogen. *International Journal of Hydrogen Energy*, 1–11.

Ripke, J. & Kopfle, J. (2017). *Direct from MIDREX*. Charlotte. Retrieved from: www.midrex.com/wp-content/uploads/Midrex_2017_DFM3QTR_FinalPrint.pdf (Last accessed: 15.02.2022)

Samsun, R., Antoni, L. & Rex, M. (2020). *Report on Mobile Fuel Cell Applications – Tracking Market Trends. Advanced Fuel Cells Technology Collaboration Programme*. Paris. Retrieved from: www.ieafuelcell.com/fileadmin/publications/2020_AFCTCP_Mobile_FC_Application_Tracking_Market_Trends_2020.pdf (Last accessed: 15.11.2021)

Soman, A., Kaur, H., Jain, H. & Ganesan, K. (2020). *India's Electric Vehicle Transition – Can Electric Mobility Support India's Sustainable Economic Recovery Post COVID-19?* Council on Energy, Environment and Water, New Delhi.

WEF (2020). *Clean Skies for Tomorrow – Sustainable Aviation Fuels as a Pathway to Net Zero Emissions*. World Economic Forum, Davos.

World Bank (2020). *Research and Development Expenditure*. Retrieved from: https://data.worldbank.org/indicator/GB.XPD.RSDV.GD.ZS (Last accessed: 25.11.2021)

Yadav, D. & Banerjee, R. (2020). Net energy and carbon footprint analysis of solar hydrogen production from the high-temperature electrolysis process. *Applied Energy*, 262. Retrieved from: https://doi.org/10.1016/j.apenergy.2020.114503.

Yadav, D., Guhan, A. & Biswas, T. (2021). *Greening Steel: Moving to Clean Steelmaking Using Hydrogen and Renewable Energy*. Council on Energy, Environment and Water, New Delhi.

5 Nuclear Energy in India's Sustainable Development Pathway

Past, Present, and the Future

A.T. Ajmal Khan

5.1 Introduction

According to India's Nationally Determined Contribution (NDC) in support of the Paris Agreement, India is promoting nuclear power as a safe, environmentally benign, and economically viable source of energy to meet the increasing electricity needs of the country. The NDC envisaged advanced technologies such as integral pressurised water reactor (PWR), advanced heavy water reactor, fast breeder reactor (FBR), and accelerator-driven systems in advanced nuclear fuel cycles to enhance the power generation from nuclear. However, contributions to the electricity mix from nuclear sources have not increased much. According to the Central Electricity Authority (CEA, 2021) of India, nuclear electricity contributed merely 6.8 GW (1.77 per cent) from 23 reactors to the country's overall installed capacity of more than 382 GW by April 2021. Back in 2015, the total installed capacity of nuclear power in operations was 5.8 GW. The NDCs mention additional six reactors with a capacity of 4.3 GW at different stages of commissioning and construction and describe how efforts were being made to achieve 63 GW installed capacity by 2032 if fuel supply was ensured (UNFCCC, 2016). This snapshot already encapsulates major insights into the achievements and shortcomings of India's nuclear power plans. This chapter is dedicated to unfolding the narrative behind these figures starting in the 1940s.

5.2 The Foundations of India's Nuclear Power Past

India's nuclear power programme begins with a few initial developments such as the establishment of the Tata Institute of Fundamental Research in Mumbai under the leadership of Homi Bhabha in 1945, the formation of the Atomic Energy Commission in 1948, and The Department of Atomic Energy (DAE) directly under the Prime Minister's office in 1954. This was followed by commissioning research reactors mostly with international partners. With this India slowly moved to a three-stage development of its nuclear power programme aimed at energy autonomy (Banerjee & Gupta,

DOI: 10.4324/9781003281818-5

2017). The three-stage development of nuclear energy laid the foundation for a solid nuclear power programme including indigenous reactor development. In stage one, nuclear power was developed from the natural uranium fuel in a pressurised heavy-water reactor (PHWR). In the second stage, the development was characterised by FBRs followed by the development of self-sustained Thorium-based reactors in the third stage. Simultaneously, the first research reactors CIRUS and APSARA were developed with the help of the United States, United Kingdom, and Canada, followed by the establishment of the Bhabha Atomic Research Centre at Trombay in Mumbai, Maharashtra, in 1967. Two years later, the first commercial nuclear power station of India, Tarapur Atomic Power Station (TAPS) went into operation, and subsequently PHWR installations were constructed in Kaiga (Karnataka), Kakrapar (Gujarat), Kalpakkam (Tamil Nadu), Narora (Uttar Pradesh), and Rawatbhata (Rajasthan). In the aftermath of the global nuclear isolation, as the result of the first nuclear bomb test in Pokhran (Rajasthan) during 1974, one important, though not immediate, development was the establishment of the Nuclear Power Corporation of India Limited (NPCIL) in 1987. The NPCIL was set up as a public sector undertaking with the objective to design, construct, operate, and maintain India's atomic power stations for the generation of electricity. This marked a new phase of massive nuclear expansion in India with nuclear power projects being constructed across the country. According to the IAEA Power Reactor Information System (2021), India currently has 23 operational nuclear power reactors, with a total net generating capacity of 6.8 GW. This comes to around 3 per cent of the total energy generated in India, which is very little output for a country with more than six decades of the nuclear power programme. Nevertheless, nuclear energy was and is told to be one of the major future sources of energy in India ever since the beginning of its nuclear programme.

5.3 India's Nuclear Power Present Between Projections and Reality

In the initial years, India was anticipating producing higher amounts of nuclear power. In 1964, Bhabha said:

> [T]here is little doubt that before the end of the century, atomic energy will be producing a substantial part of the power in India, and therefore practically all the addition to our power generation will come from it at that time.
>
> (Ramana, 2012)

Less than a decade later, Homi Sethna, the chairperson of the Atomic Energy Commission, predicted that India would have 4.3 GW of nuclear-generating capacity by the year 2000. However, the actual output from nuclear was significantly less as noted by the Comptroller and Auditor General of India in 1999, which stated:

> Against the targeted additional power generation of 940 MW by 1995–96, gradually increasing to 7880 MW by 2001, the actual additional generation of power under the profile as of March 1998 was nil in spite of having incurred an expenditure of 52.91 billion rupees.
>
> (Ramana, 2012)

Figure 5.1 presents the actual increase in operating nuclear capacity in India showing six significant advances between 1999 and 2021 with total capacity addition of 5 GW.

The NPCIL then targeted to produce 20 GW by 2020 and 60 GW by 2032 and these targets were reiterated in 2010 and even increased to 63 GW by 2032 in 2011 (WNA, 2021). Former Prime Minister Manmohan Singh argued for a civil nuclear cooperation agreement with the United States and confirmed that India's nuclear energy target for the year 2020 would be 20 GW (MEA, 2007). After the India–US nuclear deal of 2008, the waiver in the Nuclear Suppliers Group, and multiple nuclear agreements with the United States, Canada, Russia, France, Australia, and Japan, the DAE went as far as to project production of 470 GW of electricity by 2052 (Ramana & Suchitra, 2009). Led by the National Democratic Alliance, the new government continued projecting higher energy outputs from nuclear power. Endorsing higher energy requirements, the National Democratic Alliance government approved the construction of 10 nuclear reactors with a combined capacity of more than 7 GW. However, the actual output from nuclear did not increase much as per NPCIL 2021 data. In the year 2017, India's 22 nuclear reactors have reached a total production capacity of only 6.3 GW. Table 5.1 gives a comprehensive overview of currently operating reactors across the country and the amount of energy generated from each of them with the year in which they started commercial operation.

Among India's operational reactors, Kudankulam I and II generate the highest amount of output with a combined 2 GWe and the Tarapur project with four reactors generates 1.4 GWe. By contrast, Madras-I was shut down in January 2018 to carry out work on the end shield. The reactor was still offline in June 2023 but is considered operable while Rajasthan-1 is offline since 2005 and is considered a permanent shutdown even if not officially declared (WNA, 2024).

There are several projects under construction, as per Table 5.2, out of which seven reactors are expected to produce an additional 5.4 GWe. A major share of this would come from Kudankulam III and IV reactors. However, looking at the current progress of construction, it is unlikely that any of these reactors will start generating electricity at the given due date for commercial operation. Kalpakkam, for example, was due in 2020, Kakrapur III and IV and Rajasthan VII and VIII were due in 2022 (WNA, 2021; Power Technology, 2023).

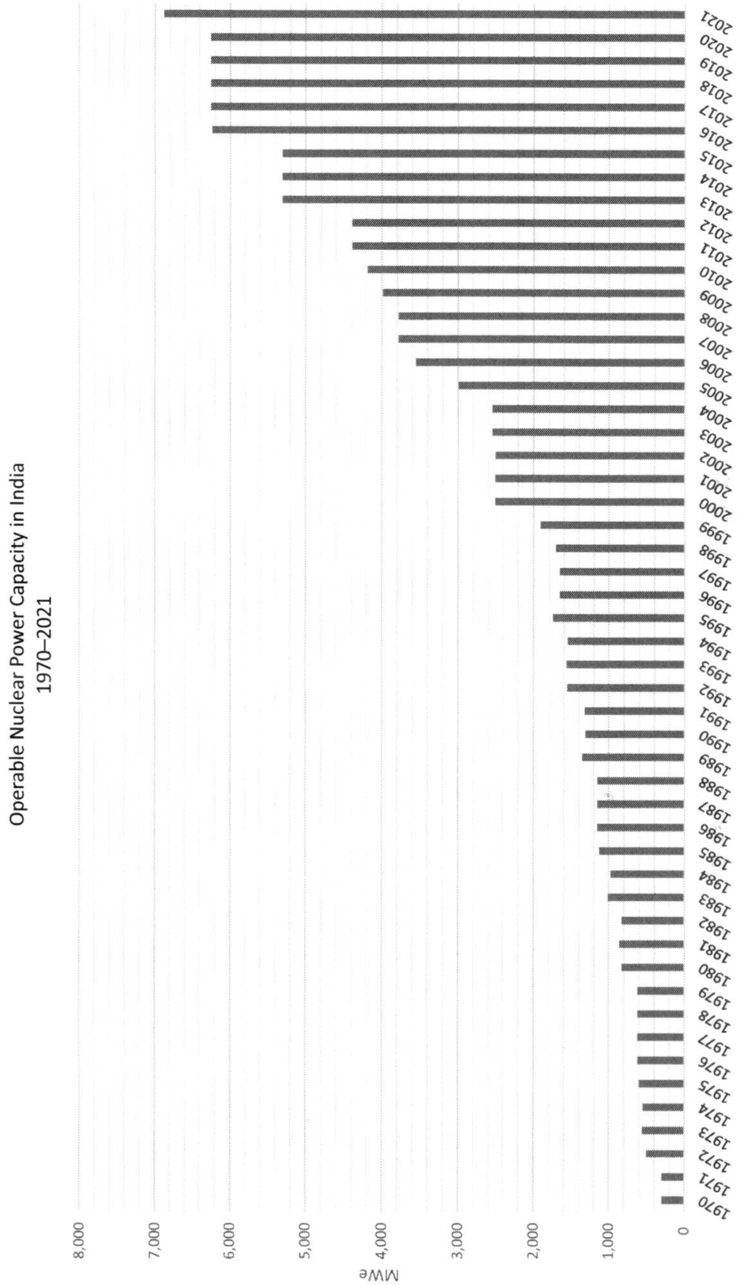

Figure 5.1 Operable nuclear power capacity in India since 1970.

Source: WNA (2021)

Table 5.1 India's operating nuclear power reactors.

S.No.	*Plant*	*Unit*	*Type*	*Capacity (MWe)*	*Date of commercial operation*
01	TAPS, Maharashtra	1	BWR	160	October 28, 1969
02	TAPS, Maharashtra	2	BWR	160	October 28, 1969
03	TAPS, Maharashtra	3	PHWR	540	August 18, 2006
04	TAPS, Maharashtra	4	PHWR	540	September 12, 2005
05	RAPS, Rajasthan	1	PHWR	100	December 16, 1973
06	RAPS, Rajasthan	2	PHWR	200	April 1, 1981
07	RAPS, Rajasthan	3	PHWR	220	June 1, 2000
08	RAPS, Rajasthan	4	PHWR	220	December 23, 2000
09	RAPS, Rajasthan	5	PHWR	220	February 4, 2010
10	RAPS, Rajasthan	6	PHWR	220	March 31, 2010
11	MAPS, Tamil Nadu	1	PHWR	220	January 27, 1984
12	MAPS, Tamil Nadu	2	PHWR	220	March 21, 1986
13	KGS, Karnataka	1	PHWR	220	November 16, 2000
14	KGS, Karnataka	2	PHWR	220	March 16, 2000
15	KGS, Karnataka	3	PHWR	220	May 6, 2007
16	KGS, Karnataka	4	PHWR	220	January 20, 2011
17	KKNPS, Tamil Nadu	1	VVER-1000 (PWR)	1,000	December 31, 2014
18	KKNPS, Tamil Nadu	2	VVER-1000 (PWR)	1,000	March 31, 2017
19	NAPS, Uttar Pradesh	1	PHWR	220	January 1,1991
20	NAPS, Uttar Pradesh	2	PHWR	220	July 1,1992
21	KAPS, Gujarat	1	PHWR	220	May 6, 1993
22	KAPS, Gujarat	2	PHWR	220	September 1,1995
23	KAPS, Gujarat	3	PHWR	220	January 10, 2021

Note: BWR, boiling water reactor; KAPS, Kakrapar Atomic Power Station; KGS, Kaiga Generating Station; KKNPS, Kudankulam Nuclear Power Station; MAPS, Madras Atomic Power Station; NAPS, Narora Atomic Power Station; PHWR, Pressurised Heavy Water Reactor; RAPS, Rajasthan Atomic Power Station; VVER, Russian acronym for Water Water Energy Reactor.

Source: NPCIL (2021)

Table 5.2 Nuclear power reactors under construction.

State	*Location*	*Project acronym*	*Capacity (MW)*	*Sanctioned cost (₹ crore)*
Gujarat	Kakrapar	KAPP 3 & 4	2 × 700	11,459*
Rajasthan	Rawatbhata	RAPP 7 & 8	2 × 700	12,320
Tamil Nadu	Kudankulam	KKNPP 3 & 4	2 × 1,000	39,849
	Kalpakkam	Prototype fast breeder reactor **	500***	5,677
Haryana	Gorakhpur	GHAVP 1 & 2	2 × 700	20,594

Note: * Under Revision. ** Project being implemented by BHAVINI. *** Excavation commenced.

Source: GoI (2019)

Table 5.3 Nuclear power reactors accorded administrative approval and financial sanction.

State	*Location*	*Project*	*Capacity (MW)*	*Sanctioned cost (₹ crore)*
Haryana	Gorakhpur	GHAVP 3 & 4	2 × 700	1,05,000
Rajasthan	Mahi-Banswara	Mahi Banswara 1 & 2	2 × 700	
		Mahi Banswara 3 & 4	2 × 700	
Karnataka	Kaiga	Kaiga 5 & 6	2 × 700	
Madhya Pradesh	Chutka	Chutka 1 & 2	2 × 700	
Tamil Nadu	Kudankulam	KKNPP 5 & 6	2 × 1,000	49,621

Source: GoI (2019)

Table 5.4 Nuclear power sites accorded "in-principle" approval.

State	*Site*	*Capacity (MW)*	*In cooperation with*
Maharashtra	Jaitapur	6 × 1,650	France
Andhra Pradesh	Kovvada	6 × 1,208	United States
Gujarat	Chhaya Mithi Virdi	6 × 1,000*	
West Bengal	Haripur	6 × 1,000*	Russian Federation
Madhya Pradesh	Bhimpur	4 × 700	Indigenous PHWR

Note: *Nominal capacity.

Source: GoI (2019)

The DAE (2017) and NPCIL are planning many more reactors to be constructed in the future. Among them, one set of reactors has been accorded the administrative approval and financial sanction (Table 5.3), and another set has received in-principle approval (Table 5.4). When the 12 reactors are accorded administrative approval and financial sanction, a total of estimated 9 GWe capacity would be reached. Out of the five sites that are accorded "in-principle" approval, a gigantic 32.5 GWe capacity addition would be made.

The Jaitapur Nuclear Power project was announced in 2010 in the Konkan region of the state of Maharashtra. With a planned capacity of 9.9 GW, it would be one of the world's biggest nuclear power stations. The project was planned to be taken up in collaboration with the French nuclear giant, Areva, which was later transferred to French electric utility company, Électricité de France. Shortly after its announcement, protests started against the Jaitapur project and escalated during the land acquisition phase (Deshpande, 2017). While the land has been acquired for the project, nothing has moved at the project site since then. A similar case is the Bhimpur project in the state of Madhya Pradesh, which was proposed in 1982 with a capacity of 2.8 GW. However, the project did not progress beyond compiling data and checking the availability of water for the project (Shastri, 2017). Another project proposed for Mithi Virdi in the Bhavnagar district of Gujarat had to shift to Kovvada in Andhra Pradesh due to delay in land acquisition and continued local protests (Dhar, 2017). While land acquisition in Kovvada is partially complete, the projected 6 GW out of the Mithi Virdi project listed in Table 5.4 needs to be deducted from the projections. Finally, the 6 GW project proposed in Haripur in the state of West Bengal was opposed by locals and the state government. As a result, the project was scrapped and the government said that it will not allow any nuclear power projects within the state. The project was put on temporary hold and is currently planned to be revived (S.C., 2011). Doshi (2016), Rao (2017), and Sanjv (2018) describe these anti-nuclear protests in India further.

Looking at the nuclear power projections and the actual power produced from nuclear in India, there is a big mismatch between projections and actual output. Studies that looked into this puzzle, such as Bhargava (1992), Ramana (2012, 2013), Grover (2014), Bidwai (2014), and Khan (2019), have presented a set of reasons why India could not produce a higher amount of power from nuclear sources of energy. They have pointed at the technical implausibility of DAE's plans, the inability of the nuclear establishment to learn from past mistakes, the uneconomical nature of nuclear power, and local opposition. They argue that the history of the development of nuclear power projects in India carries high improbability for the exponential growth of future nuclear power generation in the country as projected by DAE or NPCIL. The Kudankulam Nuclear Power Project in Tamil Nadu, for example, took more than 35 years before it started generating any power, and reactors were shut down frequently during the initial years of operation. The Comptroller and Auditor General of India report in 2017 notes:

> There were major delays in the start of commercial operations of Units I and II by 86 and 101 months respectively due to delayed supply of equipment/working documents by overseas collaborating partner, changes in design, additional works, erection delays, etc.

The report also highlights how project costs for Kudankulam Units I and II increased significantly from INR 13,171 crore (US$1.8 million) to INR 22,462 crore (US$3 million) in 2014.

5.4 The Future of Nuclear Energy in India

The above section has shown that nuclear energy projections never got anywhere near their targets. Against this backdrop, to assess the future of nuclear energy in India, it is important to look at whether India's nuclear establishment is capable of delivering higher nuclear energy outputs in the future. This should also be looked at in the context of climate change where nuclear energy is considered to be one of the future sources of energy (Moniz, 2003; Ramana, 2010, 2016; Sovacool &Cooper, 2008; Socolow & Glaser, 2009;

IAEA, 2018; Kumar, 2019). In India, the DAE and NPCIL endorse such a perspective. The DAE, for example, argues:

> Given India's large energy needs, these commitments render nuclear power to be one of the few credible and sustainable non-fossil energy resources with the least carbon footprint that can meet the robust base load demands to drive India's future industrialisation and growth. We are, therefore, convinced that for an energy-starved country like India, nuclear power offers a reliable and environmentally-sensitive answer to climate change.
>
> (DAE, 2019)

With regard to sustainable sources of energy, nuclear power has to endure the comparison with renewable sources of energy, which have started to change the equation in fundamental ways. In India, new renewable sources, that is, without traditional large hydropower plants, have become the fastest-growing sector. Since the year 2002, installed capacity in new renewables has increased by a factor of 59. Nuclear power, by contrast, had a growth factor of 2.5 only (CEA, 2011, 2021). In terms of electricity generation, the contribution from new renewable sources of energy more than doubled from 65.1 TWh in 2015 to 151.2 TWh in 2020 (BP, 2021). By contrast, over the same period, nuclear generation has increased from 38.3 TWh in 2015 to merely 44.6 TWh in 2020 (see Figure 5.2).

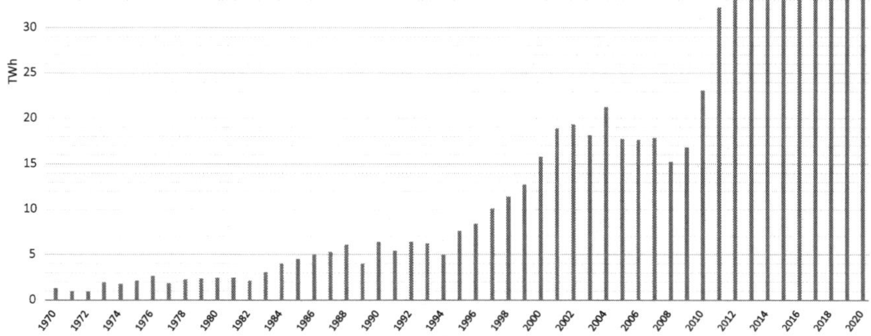

Figure 5.2 Nuclear power generation in India, 1970–2020.

Source: BP (2021)

Solar and wind energy sources together generated nearly three times as much electricity, compared to nuclear energy during the 2020–2021 fiscal year. The divergence of contributions from the renewable energy sector vis-à-vis nuclear energy is expected to further increase drastically in the coming years and decades. The International Energy Agency foresees explosive growth in solar energy followed by a somewhat more modest increase in wind energy, but relatively minuscule levels of growth in nuclear power (IEA, 2021). Studies involving modelling the grid in India even suggest that wind and solar energy "could meet 80% of anticipated 2040 power demand supplanting the country's current reliance on coal" (Lu et al., 2020). With that, there seems to be a strong economic and practical logic for rapid renewable energy expansion in India.

The nuclear sector's performance over the past year has been a continuation of earlier trends, most importantly in the form of construction delays and cost overruns. Of the six reactor projects under construction, four are delayed. The PHWR that started operating in Kakrapar was to be commissioned in 2015. The two PHWRs under construction at Rajasthan were to be commissioned in late 2016. As of December 2023, the anticipated date of commissioning of Kakrapar-4 is March 2024 while Rajasthan-7 and -8 are at an advanced stage of commissioning (MoSPI, 2023; NEI, 2023).

Finally, the prototype fast breeder reactor at the MAPS in Kalpakkam continues to maintain its status as the most delayed project. Looking at the past and the current energy generation scenarios from nuclear power, it is unlikely that even half of the energy projections are met on time. The policy elites of nuclear energy in India and the nuclear establishment have not paid much attention to this tremendous mismatch. This elite and the exclusive nature of the nuclear power industry renders it rather easy for the nuclear establishment to draw gigantic future projections as has been the case ever since the beginning of India's nuclear energy programme. By contrast, if nuclear energy was to contribute higher outputs, the energy establishment would have to pay more serious attention to the fundamental flaws of its undertakings. This includes the technical implausibility of DAE plans, the long delays in materialising a project, the uneconomical nature of the current projects, and the local protests. Even then, given the current phase of development, India's nuclear sector will only contribute slowly and relatively small proportions to total power generation.

5.5 Conclusion

Though India has had an early nuclear power programme, its output remained far behind what had been planned and many studies have highlighted why the energy output remained insignificant over the years. Electricity generated from nuclear energy continues to be around 3 per cent of the country's total power generation. Looking at the current phase of the development of new projects under construction, proposed projects, future output projections, and past experiences of the projects make it difficult to assume that nuclear power will contribute a higher share of energy in the near future. The increasing importance of renewable sources of energy, particularly solar and wind, suggests that in all likelihood nuclear energy will remain insignificant in India. Against all odds, nuclear power could only slowly increase its contributions if the energy establishment rightly addressed its various challenges.

References

Banerjee, S. & Gupta, H.P. (2017). The evolution of the Indian nuclear power programme. *Progress in Nuclear Energy*, 101: 4–18.

Bhargava, G.S. (1992). Nuclear power in India. The costs of independence. *Energy Policy*, 20(8): 735–743.

Bidwai, P. (2014). *Between Ambition and Reality: India's Nuclear Power Programme*. 07.04.2014. Retrieved from: www.boell.de/en/2014/03/21/between-ambition-and-reality-indias-nuclear-power-programme (Last accessed: 06.10.2021)

BP (2021). *BP Statistical Review of World Energy – All Data, 1965–2020*. British Petroleum. Retrieved from: www.bp.com/en/global/corporate/energy-economics/statistical-review-of-world-energy.html (Last accessed: 09.10.2021)

CEA (2011). *Annual Report 2010–11*. Central Electricity Authority. June 2011. Retrieved from: https://cea.nic.in/wp-content/uploads/2020/03/annual_report-2011.pdf (Last accessed: 21.05.2021)

CEA (2021). *Executive Summary on Power Sector*. Central Electricity Authority. March 2021. Retrieved from: https://cea.nic.in/wp-content/uploads/executive/2021/03/executive.pdf (Last accessed: 21.05.2021)

DAE (2017). *Press Release on Cabinet Decision to Transform Domestic Nuclear Industry*. Department of Atomic Energy, New Delhi. 17.05.2017. Retrieved from: https://dae.gov.in/node/973 (Last accessed: 06.10.2021)

DAE (2019). *Myths and Facts*. Department of Atomic Energy, Government of India. Retrieved from: https://dae.gov.in/node/926 (Last accessed: 08.10.2021)

Deshpande, A. (2017). Jaitapur to witness anti-nuclear protest again. *The Hindu*. 02.01.2017. Retrieved from: www.thehindu.com/news/cities/mumbai/Jaitapur-to-witness-anti-nuclear-plant-protest-again/article16974634.ece (Last accessed: 06.10.2021)

Dhar, D. (2017). The Mithivirdi movement – the untold story of a struggle against a nuclear power plant in Gujarat. *The Wire*. 30.06.2017. Retrieved from: https://thewire.in/economy/mithivirdi-movement-gujarat (Last accessed: 06.10.2021)

Doshi, V. (2016). The lonely struggle for India's anti-nuclear protesters. *The Guardian*. 06.06.2016. Retrieved from: www.theguardian.com/global-development/2016/jun/06/lonely-struggle-india-anti-nuclear-protesters-tamil-nadu-kudankulam-idinthakarai (Last accessed: 06.10.2021)

GoI (2019). *Proposals for New Atomic Power Plants*. Department of Atomic Energy, Government of India. 03.01.2019. Retrieved from: https://pib.gov.in/Pressreleaseshare.aspx?PRID=1558481 (Last accessed: 06.10.2021)

Grover, R.B. (2014). *Nuclear Power for India's Energy Security: External and Internal Challenges*. 13th Volume of India's National Security Annual Review Published by Routledge, India.

IAEA (2018). *Climate Change and Nuclear Power*. International Atomic Energy Agency, Vienna.

IAEA (2021). *Power Reactor Information System*. International Atomic Energy Agency, Vienna. Retrieved from: https://pris.iaea.org/PRIS/CountryStatistics/CountryDetails.aspx?current=IN (Last accessed: 03.07.2021)

IEA (2021). *India Energy Outlook 2021*. International Energy Agency, Paris. February 2021. Retrieved from: www.iea.org/reports/india-energy-outlook-2021 (Last accessed: 12.03.2021)

Khan, A. (2019). Interrogating State, Nuclear Energy and Social Movements in India. A Study of People's Movements in Kudankulam and Jaitapure (Unpublished PhD thesis). Tata Institute of Social Sciences (TISS), Mumbai.

Kumar, N. (2019). India's domestic climate policy is fragmented and lacks clarity. *Economic and Political Weekly*. Retrieved from: www.epw.in/engage/article/indias-domestic-climate-policy-fragmented-lacks-clarity (Last accessed: 06.10.2021)

Lu, T., Sherman, P., Chen, X., Chen, S., Lu, X. & McElroy, M. (2020). India's potential for integrating solar and on-and offshore wind power into its energy system. *Nature Communications*, 11(1): 1–10.

MEA (2007). *PM's Statement in the Lok Sabha on Civil Nuclear Energy Cooperation with the United States*. Ministry of External Affairs, Government of India. 13.08.2007. Retrieved from: www.mea.gov.in/Speeches-Statements.htm?dtl/1924/PMs+statement+in+the+Lok+Sabha+on+Civil+Nuclear+Energy+Cooperation+ (Last accessed: 07.10.2021)

Moniz, E.J. (2003). *The Future of Nuclear Power – An Interdisciplinary MIT Study (No. INIS- XA-653)*, Massachusetts Institute of Technology, Cambridge, MA.

MoSPI (2023). *Project Implementation Status Report of Central Sector Projects Costing Rs. 150 Crore & Above*. Ministry of Statistics and Programme Implementation, Government of India. July–September, 2023. Retrieved from: http://www.cspm.gov.in/english/qr/QPSR_2nd_QTR_2023-24.pdf (Last accessed: 22.01.2024)

NEI (2023). Hot conditioning completed at Rajasthan 7. *Nuclear Engineering International Magazine*. Retrieved from: https://www.neimagazine.com/news/newshot-conditioning-completed-at-rajasthan-7-11347730 (Last accessed: 22.01.2024)

NPCIL (2021). *Nuclear Power Generation (2010–11 to 2021–22)*. Nuclear Power Corporation of India Limited. Retrieved from: www.npcil.nic.in/content/302_1_AllPlants.aspx (Last accessed: 07.10.2021)

Power Technology (2023). *Kudankulam Nuclear Power Plant, Tamil Nadu, India*. Retrieved from: https://www.power-technology.com/projects/kudankulam-nuclear-power-plant/?cf-view (Last accessed: 21.01.2024)

Ramana, M.V. (2010). *The Future of Nuclear Power in India*. Centre for the Advanced Study of India, University of Pennsylvania. 04.01.2010. Retrieved from: https://casi.sas.upenn.edu/iit/ramana (Last accessed: 06.10.2021)

Ramana, M.V. (2012). *The Power of Promise: Examining Nuclear Energy in India*. Penguin Random House India, Gurgaon.

Ramana, M.V. (2013). Why India's electricity is likely to remain in short supply: The economics of nuclear power. *Bulletin of the Atomic Scientists*, 69(6): 67–78.

Ramana, M.V. (2016). Nuclear Salesmen in Paris. Wishful Thinking and Economic Realities. *Economic and Political Weekly*, 51(1). Retrieved from: www.epw.in/node/146260/pdf (Last accessed: 13.07.2021)

Ramana, M.V. & Suchitra, J.Y. (2009). Slowed and stunted: Plutonium accounting and the growth of fast breeder reactors in India. *Energy Policy*, 37(12): 5028–5036.

Rao, K.S. (2017). Opposition to Kovvada N-plant mounts. *The Hindu*. 19.01.2017. Retrieved from: www.thehindu.com/news/national/andhra-pradesh/Opposition-to-Kovvada-N-plant-mounts/article17056221.ece (Last accessed: 06.10.2021)

Sanjv, D. (2018). Karnataka: Activist Protest Kaiga expansion. *Bangalore Mirror*. 14.12.2018. Retrieved from: https://bangaloremirror.indiatimes.com/news/state/karnataka-activists-protest-kaiga-expansion/articleshow/67083089.cms (Last accessed: 06.10.2021)

S.C. (2011). West Bengal government rules out Haripur nuclear project. *The Hindu*. 17.08.2011. Retrieved from: www.thehindu.com/news/national/other-states/West-Bengal-government-rules-out-Haripur-nuclear-project/article13658088.ece (Last accessed: 08.10.2021)

Shastri, P. (2017). Fate of N-power plants hangs in balance in MP. *Hindustan Times*. 02.01.2017. Retrieved from: www.hindustantimes.com/indore/fate-of-n-power-plants-hangs-in-balance-in-mp/story-Y6hqyGCmpzOBg37USibPkI.html (Last accessed: 08.10.2021)
Socolow, R.H. & Glaser, A. (2009). Balancing risks: Nuclear energy & climate change. *Daedalus*, 138(4): 31–44.
Sovacool, B.K. & Cooper, C. (2008). Nuclear nonsense: Why nuclear power is no answer to climate change and the world's post-Kyoto energy challenges. *William & Mary Environmental Law and Policy Review*, 33(1).
UNFCCC (2016). *India's Intended Nationally Determined Contributions*. Government of India. Retrieved from: https://www4.unfccc.int/sites/NDCStaging/pages/Party.aspx?party=IND (Last accessed: 06.10.2021)
WNA (2021). *Nuclear Power in India*. World Nuclear Association. Retrieved from: www.world-nuclear.org/information-library/country-profiles/countries-g-n/india.aspx (Last accessed: 07.10.2021)
WNA (2024). Reactor database. *World Nuclear Association*. Retrieved from: https://www.world-nuclear.org/reactor/default.aspx/RAJASTHAN-1 (Last accessed: 22.01.2024)

6 Solar Energy in India

Achievements and Adoption Challenges

Parvej

6.1 Introduction

Solar energy is recognised as a clean form of energy as it does not emit the polluting greenhouse gases associated with the use of fossil fuels. India has set the target to achieve 175 GW of renewable energy by 2022, of which 100 GW from solar energy alone (PIB, 2019). At 21st Conference of the Parties in Paris, in 2015, India also decided to achieve 40 per cent (450 GW) of its power generation from non-fossil sources of energy by 2030 as part of its Nationally Determined Contributions (NDCs) (UNFCCC, 2015).

Accordingly, central and state governments have formulated policies to enhance the adoption of renewable sources of energy and especially solar energy. With Prime Minister Modi's announcement at 26th Conference of the Parties in Glasgow, in 2021, of 50 per cent (roughly 500 GW) of total installed electricity capacity through "clean" power generation by 2030 will further set the ambitious targets for solar energy to achieve by India (PIB, 2021).

The target is to achieve 100 GW of grid-connected solar energy by 2022, which includes 60 GW through ground-mounted solar power and 40 GW through rooftop solar energy. India has the installed capacity of 60.81 GW of solar power in December 2022, and the backlog of around 36 GW to achieve the target of 100 GW by 2022 has not been caught up (MNRE, 2023a). The installed capacity of rooftop solar power is around 8 GW, while ground-mounted solar power is 51.425 GW in December 2022 (MNRE, 2023a). This signifies that the highest shortfall in solar power is in the rooftop solar segment in India. This shows that there is a lesser adoption of solar power by the households or people at large scale in India. The ground-mounted solar power in India, which comprises of utility scale solar power mostly, also has a backlog of more than 8 GW, which needs to be achieved with enhanced targets in coming years.

The socio-technical transition in clean energy is driven by the accelerated change for low-carbon energy system and the availability of affordable and large-scale solar photovoltaic (PV) technology (Sareen, 2021). There

DOI: 10.4324/9781003281818-6

are various challenges to the large-scale adoption of solar energy in India. Though the price of solar panels has reduced significantly, with \$1.2/W in 2011 to \$0.2/W in 2020 for the Bloomberg monocrystalline solar panels at 2015 US dollar prices, the technology is still perceived to be expensive (Dobrotkova et al., 2018; IEA, 2023). Nevertheless, various states, including Rajasthan, Gujarat, Karnataka, and Tamil Nadu, have performed particularly well in the capacity additions of solar power generation. States like Uttarakhand, Odisha, and Haryana, on the other hand, have not kept pace with that (MNRE, 2021d). This comparative study exemplifies and specifies the reasons behind low solar PV adoption through engagement with four communities in the state of Haryana.

An exploratory research exercise has been conducted to understand the challenges of solar energy adoption. This involved a critical review of the selected existing literature available in the form of government reports, research articles, and grey literature reports of various international and domestic non-governmental organisations. Apart from this, a qualitative study using ethnographic methods was conducted in two villages and two urban areas in two districts of the state of Haryana. The study features one village with a high number of solar-equipped rooftops contrasted with a village of similar size with only very few solar PV systems installed in another district. Similarly, the urban areas were considered for users and non-users of solar energy and provide the basis for reflecting upon aspects of the rural–urban divide. This research was conducted to better understand why a particular rural community has not opted for large-scale solar energy installations, whereas another village under the same regulatory conditions of the state of Haryana has relatively more households equipped with solar panels. For this purpose, a total 83 on-field semi-structured interviews were conducted with solar PV users (40), non-users (23), DISCOM officials (5), HAREDA officials (3), TERI employees (2), one International Solar Alliance representative (1), academics (2), private players (3), solar park developers (3), and medical college staff (1). These interviews resulted in detailed conversations about the perception of solar power at the respective locations and the specific reasons for higher or lesser adoption of solar energy technology in villages.

Section 6.2 covers the background on solar energy across the world. Section 6.3 discusses the solar energy potential in India. Section 6.4 focuses on the administrative structure of solar energy-related institutions and policy initiatives in India. Section 6.5 discusses the installed capacity of solar power in India. Section 6.6 covers the installation of solar energy across the state of Haryana and presents the results of the local study. Section 6.7 puts solar energy adoption challenges into the bigger picture and Section 8.8 concludes the chapter with a discussion about policy implications.

6.2 Solar Energy – the Main Source of Renewable Energy Worldwide

Solar energy is available in abundance on earth. In recent years, solar energy has emerged as one of the primary sources of renewable energy, due to technological innovation, investment, and the general push towards a clean energy transition. Solar energy can be harnessed and converted into usable forms such as electricity and heat using technologies such as solar PV or concentrated solar power.

In general, solar has advantages over other forms of energy, such as wind and biomass, due to its high availability, lesser capital cost, and low maintenance cost. Despite the day-night constraints, sun is a fairly constant source of energy and there is a fixed average amount of solar radiation in a place over the one-year cycle. However, the intensity of solar radiation depends on various factors such as location parameters, time of day, and weather (OEERE, 2020).

Globally, renewable energy installations have been expanding in the light of meeting the carbon emission reduction goals of various countries. Solar energy is abundant on earth. However, the global horizontal irradiation, which denotes the solar potential of an area, varies across regions. It is one parameter to assess the suitability of a region to harness solar energy. Higher solar irradiation correlates with higher solar potential of a specific location.

Countries and regions such as Chile, Australia, North and South Africa, Middle East Asia, and Central America have very high solar potential (OECD/UN, 2018). These regions have global horizontal irradiation above 6 kWh/m^2, which puts them at an advantage in terms of harnessing solar energy. Similarly, countries like Brazil, India, China, and United States have significant solar potential (World Bank Group, 2020).

As represented in Figure 6.1, the global cumulative capacity of solar PV has been increasing consistently from 39 GW in 2010 to 935 GW in 2021. Globally, the annual addition in solar PV capacity has increased with more than 100 GW over the past five years with addition of solar PV capacity by 175 GW in 2021 (REN21, 2022). This shows that solar power is growing rapidly the world over. Similarly, India has witnessed significant addition of installed capacity of solar PV from 6.8 GW in 2016 to 63.3 GW by December 2022 (MNRE, 2023a; NPP, 2022).

6.3 Solar Energy Potential in India

The solar energy potential relates to the usable solar energy radiated on a specific area. India's total estimated solar energy potential is approximately 750 GW, assuming 3 per cent of India's waste land to be covered by solar PV (MNRE, 2022). Being geographically and area-wise diverse, the states of India have a varied range of solar energy potential. In most parts of the country, India's horizontal irradiation has a potential of around 4.5 kWh/m^2,

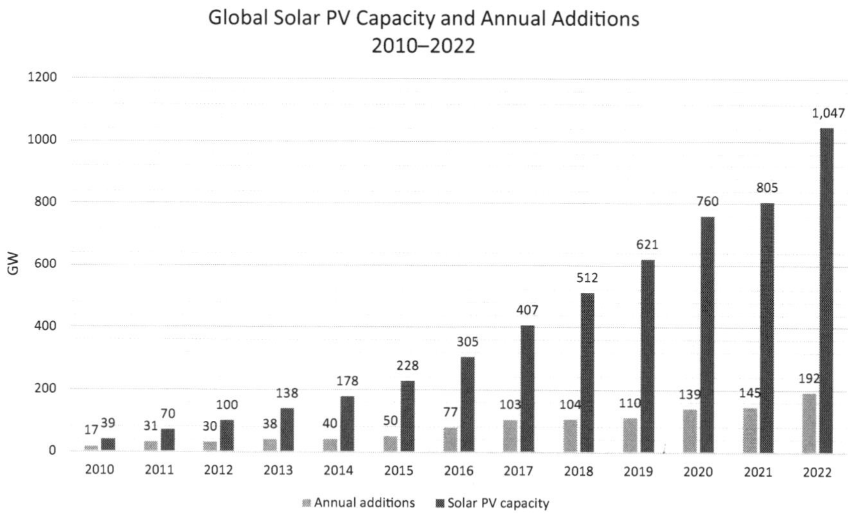

Figure 6.1 Global solar PV capacity and annual additions, 2010–2022.
Source: REN21 (2022), IRENA (2023)

whereas the western and southern states can go up to 5 kWh/m^2 and above making them potentially more favourable for solar energy installations.

As represented in Figure 6.2, states such as Rajasthan, Jammu and Kashmir, Maharashtra, and Madhya Pradesh have particularly high solar energy potential. With high solar irradiation and limited domestic reserves of fossil fuels, India is in a prime position to develop the use of renewable energies and solar energy in particular.

Figure 6.3 depicts the declining trend of solar power tariff in India in last around seven years. It has decreased from INR 6.46/kWh in 2014 to INR 1.99/kWh in 2021. The declining prices of solar power tariff show the positive sign for large-scale adoption of solar energy in India.

6.4 Administrative Structuration of Solar Energy Utilisation and Policy Initiatives in India

The administrative organisation of solar energy in India, as illustrated in Table 6.1, is operative at three levels: central government, state government, and district administration.

6.4.1 Central Level

At the central government level, the Ministry of New and Renewable Energy (MNRE) is the key agency in charge of solar energy. MNRE is the nodal

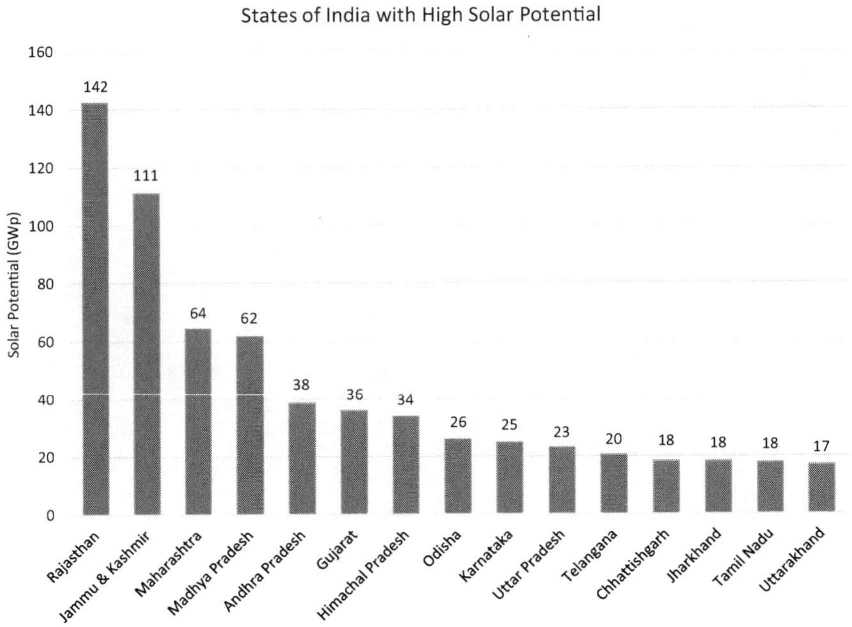

Figure 6.2 Indian states with highest solar energy potential.
Source: MNRE (2022)

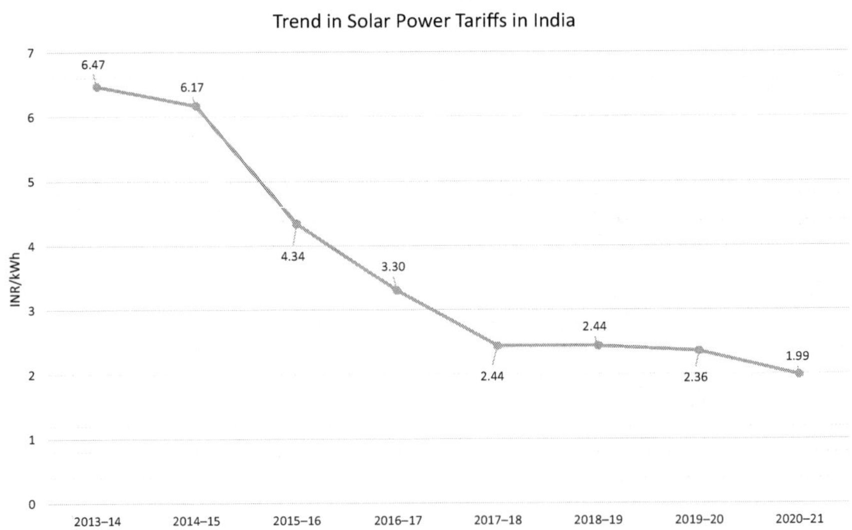

Figure 6.3 Trend of solar power tariff in India since 2013.
Source: MNRE (2021d, 2022)

Table 6.1 Administrative organisation of solar energy in India.

Central agencies (MNRE)			*State agencies*		*District agencies*
IREDA	NISE	SECI	REDA	DISCOMs	DISCOMs

agency of the Government of India related to all renewable energy projects (MNRE, 2021b). It aims at deploying renewable energies by framing policies and schemes to achieve government targets. MNRE has various institutions under its jurisdiction and administration such as Solar Energy Corporation of India (SECI), Indian Renewable Energy Development Agency Limited (IREDA), and National Institute of Solar Energy (NISE), which deal with and assist in solar energy projects. SECI was established in 2011 as a state-owned company under MNRE to facilitate the Jawaharlal Nehru National Solar Mission (SECI, 2020). The National Solar Mission was launched in 2010 to achieve 20 GW of solar energy in India by 2022. In 2015, the target was revised and upgraded to 100 GW of solar energy by 2022. IREDA, a government enterprise under MNRE, was set up already in 1987 as a non-banking financial institution to promote and facilitate the financial assistance for projects related to renewable energies, energy efficiency, and energy conservation in India (IREDA, 2020). NISE is an autonomous institution set up under the MNRE as an apex research and development body dedicated to solar energy. It aims at assisting the MNRE in the implementation of the National Solar Mission and in conducting research- and technology-related works in the field of solar energy (NISE, 2020).

6.4.2 *State and District Levels*

At the state level, Renewable Energy Development Authority (REDA) of each state acts as nodal agencies of the MNRE. This agency has the mandate to facilitate the wider adoption of renewable energies in their respective states. State power distribution companies (DISCOMs) have been mandated to act as the implementing agencies for grid-level solar rooftop programmes. The central finance assistance in terms of subsidy is provided through DISCOMs. At the district level, the subsidy is channelled through DISCOMs to the residential electricity consumer.

The MNRE has rolled out policies and schemes to enhance the adoption of solar power in the country. The following subsection will provide an overview of the major active schemes for enhancing solar power in India.

6.4.3 *Solar Power Policy Initiatives*

As part of its updated NDCs, the Government of India decided to achieve 50 per cent of its cumulative installed electricity capacity from non-fossil fuel sources of energy by 2030 (PIB, 2021). The Jawaharlal Nehru National Solar Mission of 2010 achieved its initial target of 20 GW already in 2016 as a result of ambitious sub-national targets and aggressive actions in states such as Karnataka, Rajasthan, and Telangana. Accordingly, the National Solar Mission target for 2022 was increased to 100 GW with 60 GW from ground-mounted solar power generation and 40 GW from solar rooftop PV. In 2019, the MNRE and state governments reformulated solar power policies offering new subsidies to enhance the solar ecosystem and to achieve the ambitious target of 100 GW solar power by 2022.

Solar energy can be utilised through three systems, distinguishing on-grid, off-grid, and hybrid technologies, depending on their linkage with the power grid and battery storage capacity available for solar power. The predominant part of government schemes in support of solar utilisation, however, is grid-based as per the following overview of public policy initiatives.

6.4.3.1 *Grid-Connected Solar Parks and Ultra Mega Solar Power Projects*

The solar park scheme was launched by the MNRE in 2014. Initially, at least 25 solar parks and ultra-mega solar parks of a total capacity of 20 GW were proposed to be set up. In 2017, the targeted capacity was enhanced to 40 GW to be achieved by 2021–2022. Under this scheme, the capacity of each solar park shall be a minimum of 500 MW, though smaller solar parks are also considered in case of limited availability of non-agricultural land and difficult terrains. These solar parks would be developed in collaboration with state governments and agencies, central public sector undertakings, and private entrepreneurs. The central government provides financial assistance of INR 2.5 million (approximately US$33,500) per solar park, and INR 2 million per MW or 30 per cent of the project cost, whichever is lower if targets are achieved within time (MNRE, 2021c).

6.4.3.2 *Grid-Connected Solar Rooftop Program*

The Solar Rooftop Program aims to achieve a cumulative capacity of 40 GW from solar rooftop projects in India by 2022. It has two components: Component A relates to the Central Financial Assistance (CFA) for the residential sector with a cumulative capacity of 4 GW. The CFA covers 40 per cent for capacity of up to 3 kW, and 20 per cent for capacity between 3 kW and 10 kW. The programme also provides CFA of 20 per cent for housing societies installing rooftop solar PV with an overall capacity of up to 500 kW. The implementing agencies are State Power Distribution Companies (DISCOMs). Component B provides an incentive to the DISCOMs for achieving aggregate solar power capacity of up to 18 GW (MNRE, 2021c).

6.4.3.3 *Pradhan Mantri Kisan Urja Suraksha Evam Utthaan Mahabhiyan*

The Pradhan Mantri Kisan Urja Suraksha Evam Utthaan Mahabhiyan scheme aims to add a capacity of 25.7 GW through solar and other renewable

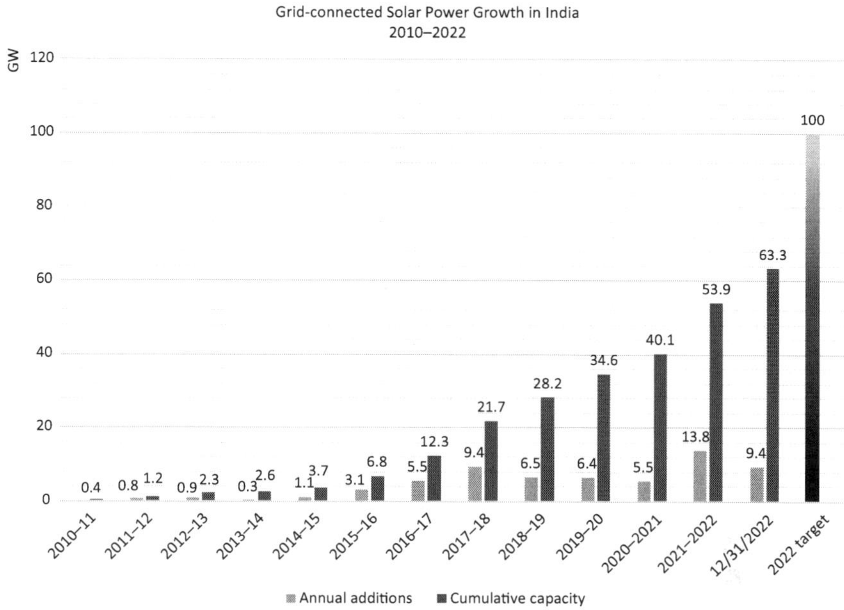

Figure 6.4 Year-wise achievement of grid-connected solar power in India, 2010–2022, by December 31, 2022.

Source: MNRE (2023b); NPP (2022)

energies by 2022. For this purpose, central financial support of INR 34,422 crore (more than US$46 million) has been allocated. The scheme consists of three components. Component A includes 10,000 MW of decentralised ground-mounted grid-connected renewable power plants. Each individual power plant will have a capacity of up to 2 MW. Component B includes the setting up of 17.5 lakh stand-alone solar-powered agricultural pumps with individual pump capacity of up to 7.5 HP. This off-grid solution operates independently of the power grid and is generally useful in remote locations. Component C includes the solarisation of existing 10 lakh grid-connected agricultural pumps with individual pump capacity of up to 7.5 HP (MNRE, 2021c).

6.5 Installed Solar Power Capacity in India

The National Solar Mission mandated for an amendment of the 2006 National Tariff Policy, making it obligatory for the states to purchase solar power at 3 per cent of total consumption by 2022 (MNRE, 2021f). On the production side, India has set itself the ambitious target of achieving 100 GW of installed solar energy by 2022 (PIB, 2019). Figure 6.4 shows the year-wise capacity additions of solar power in India connected to the power grid. Starting out in 2010, India's solar capacity was around 11.4 MW. One decade

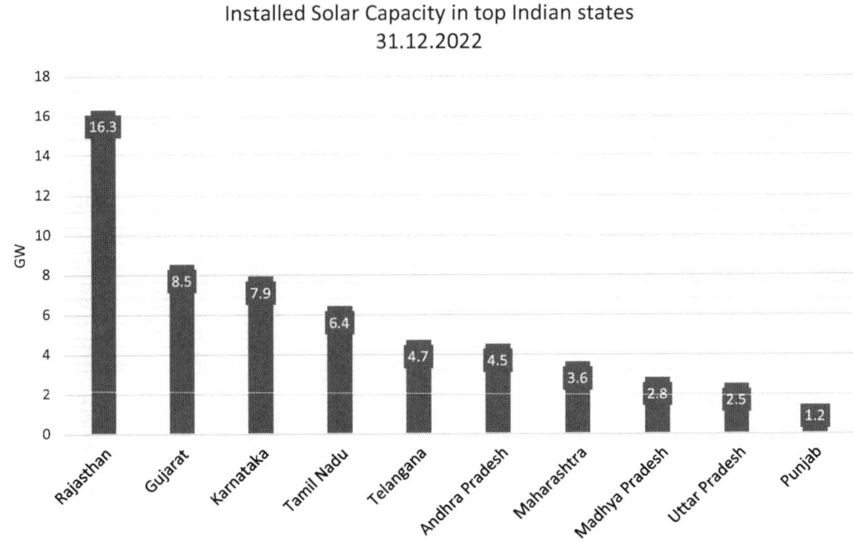

Figure 6.5 Installed solar power capacity of states in India on December 31, 2022.
Source: MNRE (2023a)

later, the cumulative installed capacity of solar power has reached 63.3 GW by December 31, 2022 (NPP, 2022).

Corresponding to centre-level policy formulation, many states have issued their solar energy policies. With the launch of the National Solar Mission in 2010, many state governments took the initiative to push for solar power installations. States such as Karnataka and Rajasthan lead in having notified their solar power policy as early as 2011. They are followed shortly by states, including Madhya Pradesh (2012), Tamil Nadu (2012), Chhattisgarh (2012), Odisha (2013), Jammu and Kashmir (2013), Uttar Pradesh (2013), Uttarakhand (2013), Kerala (2013), Himachal Pradesh (2014), and Haryana (2014), notifying their policies to expand the solar power utilisation within their respective region and legal entity. After that, many other states such as Andhra Pradesh (2015), Telangana (2015), Gujarat (2015), and Jharkhand (2015), among others issued their solar power policies. Figure 6.5 shows the achievements in the form of installed capacity of solar power in the states of India by December 31, 2022. The state of Rajasthan has the highest installed capacity of solar power in India. The states of India with higher solar power capacity are Rajasthan (16,340.75 MW), Gujarat (8,500.74 MW), Karnataka (7,885.57 MW), Tamil Nadu (6,412.36 MW), Telangana (4,650.93 MW), Andhra Pradesh (4,524.72 MW), Maharashtra (3,646.13 MW), Madhya Pradesh (2,774.78 MW), Uttar Pradesh (2,485.16 MW), and Punjab (1,153.21 MW).

6.6 Solar Energy in Haryana: Research Results and Discussion

The state of Haryana is situated in Northern India and surrounds the capital city New Delhi to the north, west, and south. According to the International Energy Agency's India Energy Outlook (IEA, 2021), Haryana has got the second-highest purchasing power after Delhi in entire India. In terms of population, 16.5 million people live in the rural areas of the state, while 8.8 million live in urban spaces. With more than 2,000 kWh per capita per year, the state also ranks among the top electricity consumers in the country. With a territory of more than 44,000 square kilometres, the state of Haryana has got sufficient space to produce clean electricity for its own consumption and that of the capital city of Delhi. However, the NISE assesses Haryana's solar energy potential at only 4.56 gigawatt-peak, which puts the state at the lower end of the Indian solar spectrum (MNRE, 2021d). Nevertheless, the state of Haryana issued its first solar power policy in 2014. In 2021, a new state draft solar power policy has been issued, which aims at increasing ground-mounted solar power and achieve the renewable purchase obligation of solar power for DISCOMs in the state. The policy also aims at facilitating the establishment of solar parks and solar power projects for the sale of electricity to DISCOMs (DNRE, 2021). State-wide by October 2021, the installed capacity of off-grid solar utilisation energy reached 194.18 MW, whereas the installed capacity of grid interactive solar power has reached 544.87 MW, of which 349.07 MW are rooftop solar PV and 195.8 MW are ground-mounted solar installations. So, the state of Haryana has total solar power installed capacity, both off-grid and on-grid, at 739.05 MW by October 2021 (MNRE, 2021e).

Two districts in Haryana – Sonipat and Hisar were part of this study with village A and city C in the district of Sonipat, and village B and city D in the district of Hisar. Overall, as per latest estimates of DISCOM officials (UHBVNL), the district of Hisar has adopted grid-connected solar power in significantly higher number (>30 per cent). However, when corrected by population figures, the number of installed systems evens out compared with Sonipat district as presented in Table 6.2. By contrast, what is striking is the relatively higher number of solar systems installed in the Sonipat village and the Hisar urban area compared with the corresponding entity of the other district.

In the following subsection, we will explain differences and similarities in outcomes across the four different cases. As the tabled data show, village A has higher adoption of solar energy than village B. With the recent influx of domestic migrants, the population of village A increased, reaching a total number of households of around 850 in 2019 as per the estimates of the Sarpanch – the Panchayat Head. In this village, 80–100 households have installed solar power on the roofs of their houses. Over the past 15 years, the government has acquired land in and around this village. This has led to an increase in disposable income among the locals. People are engaged

Table 6.2 On-grid solar systems in selected settlements in the districts of Sonipat and Hisar, Haryana.

On-grid solar power installations vs. population/households	*Sonipat district*	*Hisar district*
Rural	**481**	**373**
(one village)	3,699/850	4,367/890
Urban	**270**	**721**
(one city)	289,333/57,740	423,945/84,677
Total	**751**	**1,094**

Source: DCOH (2011a, 2011b)

in agricultural activities, private sector jobs, and cab services that cater to the industrial and educational institutions set up in its vicinity over the past around ten years.

Despite a comparable number of households with 4,367 inhabitants, the village in the district of Hisar has a lower adoption rate of solar energy. People in this village are primarily engaged in traditional agricultural activities. A small share of its population is working for the government and in private sector jobs. The average income level of people in this village is markedly less than people of village A, which correlates with significantly lower number of households (five) in village B with installed solar modules on the roof of their houses in the year 2019.

Households in the urban areas of Haryana – cities C and D – were approached for detailed conversations to better understand their perception of solar energy adoption in North India. The interviews included users and non-users of solar energy across cities of Haryana. Here, diverging conditions in terms of electricity supply, level of education, financial situation, and governance prevail compared to rural areas. Arguably, Hisar's urban area has more local industry, which might explain the relatively higher solar adoption in that part of the district, compared with the urban area in the district of Sonipat.

This comparative analysis is meant to support policymakers and other stakeholders in the solar energy ecosystem to overcome the challenges faced by people in rural and urban areas of Haryana. A few core insights have emerged that explain the convergence and divergence in outcomes across the studied entities.

Many factors could be identified, which can be broadly categorised as socio-economic factors, policies and political will, awareness and education, and technical aspects – energy infrastructure, electricity availability, and reliability. These factors explain the generally low adoption of solar energy in rural and urban areas in India. Generally, a different set of challenges and dynamics determine the adoption of solar power in rural areas compared to urban areas. However, these challenges shape people's perception when they consider installing solar energy systems at household level in India.

6.6.1 Socio-Economic Factors

The major barrier to the installation of a solar system for the individual households is the high capital cost. Though the price of solar panels has reduced to less than half in the past few years, people perceive it as still costly. Currently, the average cost of rooftop solar PV without government subsidy is around INR 47,000/kW (630 USD/kW) (MNRE, 2021a). A subsidy is not available for off-grid residential purposes. With government subsidy for an on-grid solar system between 10 and 40 per cent, the cost comes down to INR 28,000 to 42,300/kW (375–570 USD/kW), depending on the total installed capacity. Apart from the mere solar system, people may need battery storage which would further increase the costs. Interestingly, it is mainly people in rural areas that perceive solar systems as expensive. Interviewees in both the villages found cost the major factor for not installing a solar system. Many respondents said that with this much upfront cost, they would prefer not to have a solar system as its cost would cover for their electricity bills for years. The comparatively less disposable income in the villages explains the generally lower adoption of solar energy in rural areas. People in urban areas have higher disposable income which makes an investment in solar energy more likely.

The adoption of solar energy in agriculture is also limited by various factors. These include highly subsidised agriculture power supply[1] and spatial limitation of water availability as expressed by people in both the villages. The groundwater is not readily available at many places and its level has also gone down at places. This puts a limitation on the utility of solar water pump. The solar energy generated through the solar water pumping system is also not fully utilised, except for the duration of pumping groundwater. The solar water pumping does not operate during night-time and can operate only during sunny days, though the government of Haryana provides for 90 per cent subsidy on solar water pumping system.

A disincentive for the usage of solar energy is the no alternative of stealing electricity in on-grid solar power system. If a household opts for the on-grid solar energy, then it would not be able to steal the electricity through the hooking of cables. If it steals the electricity by hooking the grid wire through its metered connection, then the solar power generated will not be fed into the grid leading to loss. It happens so as it bypasses the electricity meter leading to no record of solar power generated and transferred into the power grid. Electricity theft is an indicator for low-income households and a criminal practice more commonly found in rural areas where DISCOM losses are higher than in urban settings. It is evident from the fact that DISCOMs operate with higher financial losses in rural areas than urban areas for residential power supply in Haryana. The current aggregate technical and commercial (AT&C) losses[2] in the state of Haryana are around 17.12 per cent (UDAY, 2023). The current national average AT&C losses for India are around 17.38 per cent (UDAY, 2023).

DISCOMs do not pay for the extra solar power generated by the household rooftop, apart from a financial incentive per unit of solar power generated.

Households consider the extra solar power generated as wastage because it is not paid at the full price. It is considered only to balance the total number of units of electricity consumed by the household from the power grid and the solar power generated into the power grid by the household.

For the off-grid solar system, high cost due to battery storage is the main barrier in its adoption by people. There is also no provision of government subsidy on off-grid solar energy in India except for the solar water pumps. Heavy appliances also do not operate on solar energy due to their limited power capacity. Most of the users and non-users say that the maintenance cost of battery storage and its replacement cost turn out to be expensive for them.

The perception of people towards the installation of solar power is also determined by the reduction in electricity bills. The users of solar energy in urban areas do notice a reduction on their electricity bills. This happens due to the adjustment of solar power generated in the electricity drawn from the power grid through the net-metering concept.[3] One user of solar energy in an urban area who has installed a 4-kW solar system says that the electricity bill of his household is nil due to net-metering.

People in rural areas, by contrast, perceive a similar scenario differently. In the village with higher solar PV adoption, the users are mainly of two categories: entrepreneurs and large landlords. Entrepreneurs include people who have the technical know-how such as electricians. These people know the market dynamics of solar energy. It is comparatively easy for them to operate the solar system as they have the installation and maintenance knowledge for solar modules and battery storage. Large landlords or rich people who have installed solar systems with higher capacity (more than 3 kW) perceive it as a profitable venture. These people rent out solar-generated electricity to their tenants or use it for other purposes such as submersible pumps. For small users, this is not profitable because there is no reduction in their electricity bill neither before nor after the installation of a solar system. It is because the off-grid solar power takes care of the consumption of electricity during power cut in the power grid during daytime or recharging of the batteries. So, there is no perceived change in their electricity bills (making them feel that solar energy is not profitable).

In case of a higher-capacity solar system, the perception of people is different. For instance, the user of a 5-kW off-grid solar power system in village A had an electricity bill of INR 2,000–2,500 (US$27–34) before installing the solar power. After the installation, the electricity bill has come down to around INR 900–1,000 (US$12–14). Similarly, the user of an 8-kW solar system used to get a bimonthly electricity bill of INR 3,000 (US$40), now gets an electricity bill of around INR 1,000. The reduction in electricity bills is evident in the case of off-grid solar power systems. Electricity bills vary, however, depending on factors such as battery storage and power load or consumption in the household.

6.6.2 *Technical Aspects: Energy Infrastructure, Electricity Availability, and Reliability*

In rural areas, the non-availability of 24 × 7 electricity is the major driver for adopting an off-grid solar PV system at household level. Limited access to grid-based electricity provides an incentive for off-grid solar power solutions, but that comes with significantly higher capital investment primarily due to battery storage needs and the lack of government subsidy for off-grid solar.

In rural area, the technically limited availability of grid-based power supply is the major hurdle for adopting an on-grid solar system. As per the villagers, the power cut happens mostly during the daytime, lasting up to 4–6 hours when the sun is actually available. Under current technological constraints, grid-connected solar power generated during a power cut cannot be utilised by the owner. As per the guidelines of the DISCOMs in Haryana, the on-grid solar system can deliver electricity only to the power grid but not to the household that provides the system (HAREDA, 2016). DISCOM officials state that it will lead to grid instability if consumers have the liberty to divert the solar power to their personal usage at their will instead of the continuous solar power supply to the power grid. This provision has been criticised and questioned by many stakeholders.

The users (in both rural and urban areas) say "*due to power cut especially during summers, we cannot utilise the solar power for our household purposes. It gets wasted as it cannot be fed into the power grid during power-cut.*" DISCOM officials say that it is possible to deploy a simple technology to utilise the solar power even during a power cut by diverting the solar power generated to the house instead of the power grid. The state DISCOMs mention grid stability as the reason for not diverting the solar system to the household usage during power cut as stated by the DISCOM official during conversation.

Solar technology is capital-intensive mainly due to the non-availability of raw materials in India, as a result of which India is largely dependent upon imports of solar energy technology primarily from China.

The state of Haryana is generating surplus electricity, which provides little incentive for the state government to invest more in solar energy (expressed by one state DISCOM officer).

Solar park developers have concerns for the viability of solar power projects in the state. As stated by one solar park developer:

> *There is no clarity on solar power policy in Haryana primarily as there have been frequent changes in provisions related to open access and captive solar power in the state. The land prices have been high leading to increased cost of the project resulting in lesser return on investment. DISCOM officials have little knowledge and clarity on the solar power policy such as net-metering leading to delay in implementation and operational aspects of the solar projects.*

They have also expressed their concerns on the problems faced by coordination from the nodal agency and DISCOMs on the implementation of solar power projects related to power evacuation or transmission. Solar park developers also mention that the DISCOMs often put forth the concern that the increased solar power would lead to lesser buying of electricity from them, making DISCOMs insecure and sceptical of higher adoption of solar power.

6.6.3 Governance and Administrative Challenges

The major administrative issue is the timely distribution of subsidies to customers. In the study, it was found that it takes years for people to receive subsidy in their bank account. The concerns relating to subsidy disbursement are the lack of funds with the authorities, a lot of paperwork, and lengthy procedures. The officials of HAREDA say that the issue of subsidy is mainly at the central or state level. It is not much of an issue at the district level. The central and state agencies do not disburse enough funds. The majority of the funds (around 75 per cent) by the MNRE are released after around 15 months of the installation of the solar system. In 2019, MNRE has changed the provision of subsidies on solar energy by directly providing it to the DISCOMs. Now, consumers can get the solar system at subsidised price from the vendor. After the clearance, these vendors will receive the subsidy directly from the DISCOM. Many vendors have expressed concerns about timely payments by DISCOM. One major challenge for solar energy vendors is to invest in advance while the subsidy payment from the DISCOM takes time to reach the vendors. This study found that the subsidy is claimed more by solar PV users in urban than in rural areas. Of the household studies in both the villages under consideration, even a single household didn't install the solar power panels at subsidised prices. On the other hand, the users in the cities installed their solar power panels at subsidised prices. In cities, people operate the solar power through on-grid solar system while users in villages operate through off-grid solar system.

This can be attributed to the limited awareness of people and limited feasibility of the on-grid solar system in rural areas. The on-grid solar system is quite less prevalent in rural areas than the urban areas. One of the determinants for this is the provision that on-grid solar system cannot be utilised for household consumption during power cut especially during daytime. The power cuts are huge during daytime in villages in Haryana.

One of the major concerns for establishing big solar projects has been the opposition to solar energy by the DISCOMs in the state.[4] The DISCOMs oppose solar energy in their power grid due to the intermittent nature of solar power (fluctuation of power generated) leading to uncertainty in their grid connectivity and functioning. An official from HAREDA says:

> *DISCOM is the biggest hurdle for expansion of solar energy in Haryana. DISCOMs have a narrow outlook for the power distribution and care more about their grid stability instead of diversifying their source*

of generation. They have concerns of PPAs and erratic nature of solar power leading to their grid instability.

The discrepancy in the billing system is one of the administrative issues on the part of the DISCOM that occurs with the net-metering system. Here, due to the lack of smart and automatic meters which enable the recording of import and export of electricity (solar and power grids), there is no proper and regular updating of electricity units maintained by DISCOM officials. This results in ambiguity in the billing system. Consumers do not have the real-time record of solar power generated and consumed reducing the incentive for the adoption of solar energy. All these concerns can be resolved through the application of advanced technology. However, DISCOMs are not proactive in their approach towards the usage of solar energy even if installing simple technology can solve these issues of billing and real-time management.

6.6.4 *Awareness and Education*

For wider adoption of any technology, there is need of enhanced level of knowledge and awareness among the stakeholders primarily the end-users. There is lack of awareness among both people and vendors about solar energy. People are not aware of the pricing, maintenance, and usage of solar panels. People end up buying low-efficiency solar modules which are a bit less costly due to limited awareness about them. Most of the users in both the villages of study did not know about the types and features of solar modules except for their capacity (kilowatt). Vendors are not experienced enough to convince people to adopt solar energy. Private players are also not able to provide proper maintenance of the solar system. Solar systems installed in public spaces like streets and parks are facing many issues due to their low-quality and the lack of regular maintenance. This study found that many of these issues occur due to centralised management of solar infrastructure and solar power accessories. Panchayat members, by contrast, who are best suited to address such concerns at the local level, do not have any control over these activities.

There is a lack of data at the district level with the authorities. HAREDA and DISCOMs do have access to data for on-grid solar energy. Private players have access to off-grid solar energy. But there is limited coordination among these players as there is no formal mechanism of data-sharing especially for off-grid solar energy at district or lower levels. This creates challenges in effective decision-making and planning about solar power.

6.6.5 *Policy and Political Will*

There is a cap on the capacity of solar system for a household through the on-grid system based on the power load of the household (DNRE, 2016). There is a limit on the subsidy based on the installed capacity of the solar system. For instance, the subsidy on grid-interactive solar system is available up to 10 kW. This limits the expansion of solar energy in the on-grid system for household purposes, including the residential compounds also. There is also a cap on the power to be bought by the DISCOM from household consumers (DNRE, 2016). The provisions in the Haryana Solar Power Policy, 2016 allows for the adjustment of solar power generated to 90 per cent of the power consumed by the household. These provisions limit the potential enhanced usage of solar energy in the state of Haryana. In the grid-interactive solar system, excess power generated is not utilised due to one-way supply to the power grid and not to the household in case of power cut. Governments can come up with a policy measure to allow the excess power supply to be utilised by the user by allowing it for other usages such as storage and self-consumption.

There have been uncertainties in the subsidy provisions. They have been changed by the government, including MNRE at times, leading to confusion among the stakeholders including dealers, private players, and consumers. This research found that many non-users in both rural and urban areas keep on waiting for higher subsidy rates by the government, resulting in delays to the adoption of solar systems.

6.7 Solar Energy Adoption Challenges: The Bigger Picture

There have been many challenges observed in the large-scale adoption of solar power in India. The high initial capital investment or upfront cost for solar projects are a major barrier in opting for the solar PV system by people in India at large scale (Tyagi et al., 2019; Karakaya & Sriwannawit, 2015; Urpelainen, 2016). Due to few performance benchmarks, banks find it difficult to assess the financial viability of solar energy-related projects (Tyagi et al., 2019). The relatively high cost of solar power plants can be attributed to the cost of solar modules plus battery storage.

Under the CAPEX model where the consumer pays the total initial cost of a solar system, even when considering subsidies provided by the government, the investment is perceived as quite expensive (Dutt, 2020). Under the OPEX/RESCO model, where the consumer does not need to pay the upfront cost, the vendor installs the solar system and sells the power to the consumer. Here, the vendor bears the risk of uncertainty in payment and return on investment (Dutt, 2020). Nevertheless, a survey conducted by Mercom India on renewable energy in India indicates that around 82 per cent of both residential and commercial respondents are either somewhat likely or highly likely to pay more for solar power (Mercom, 2014). This suggests that there are other factors apart from the cost of solar PV which affect the adoption of solar energy in India.

The lack of awareness among end-users leads to lesser adoption of solar energy (Rathore et al., 2019). This includes the benefits, operations, and government regulations regarding solar energy. There is a lack of trust in the performance of solar projects due to the limited availability of data on their performance or output (Tyagi et al., 2019). This is aggravated by the prevalence of companies with limited market experience. The Mercom India survey on renewable energy further specifies that around 42 per cent of commercial and 55 per cent of residential respondents in India are not aware of the subsidy programmes for solar systems provided by centre and state governments (Mercom, 2014). At the same time, around 87 per cent of the respondents are in favour of government subsidies for solar power systems.

The lack of clarity on rooftop solar policies by states, more specifically, creates additional barriers to the expansion of solar power in India (Rathore et al., 2019). Regular changes to policies and guidelines on subsidies and net-metering have resulted in confusion among all stakeholders including developers, dealers, power-utilities, and end-users of solar power (Tyagi et al., 2019). In effect, these changes made by MNRE and state agencies create uncertainties for solar power-based business models and their uptake. Moreover, solar rooftop connections in India take around 90 to 100 days to get operational (Mercom, 2014). This shows that the bureaucratic hurdles and red tap lead to much delay in getting solar connection, creating obstacles for the wider adoption of grid-connected solar power systems.

The high dependence on solar technology imports primarily from China and limited grid connectivity of solar power also act as barriers to large-scale adoption of solar energy in India (Goel, 2016). Around 79 per cent of the solar panels imported to India are made in China (Parikh, 2020). Due to high import taxation, this leads to the high cost of solar panels.

6.8 Conclusion and Policy Implications

The adoption of solar energy in India has increased majorly since the announcement of the National Solar Mission in 2010. The central government and state governments are allocating dedicated funds to meet the targets to achieve the climate change commitments. This study has presented and analysed various aspects such as economic, governance, and policy factors responsible for the low adoption of solar energy in rural and urban areas in the state of Haryana in India. The households in urban areas have higher adoption of solar rooftop, which is attributed to access to finance, increased awareness of solar technology, and knowledge of regulations about solar power than the people in rural areas in India, though each segment has its own challenges to achieve the enhanced solar power in India.

Solar energy technology has its limitations such as high capital cost and operations limited to daytime. This being said, various factors can be looked upon to address the specific concerns by the stakeholders such as government agencies, private players, DISCOMs, users, and non-users. The proactive approach by DISCOM can certainly help in increasing the adoption of the solar energy. People will increasingly adopt the solar energy when they realise its intended benefits.

There are various challenges in the large-scale adoption of solar energy in India. These include high capital cost, limited awareness, import of technology, and delay in solar power connection and subsidies. The perception of people towards solar energy is largely determined by the reduction in their electricity bills. In the study, it is found that for small users or poor people, solar energy is neither profitable to them nor are they aware of it. Though the price of solar panels has reduced significantly in the past one decade, still it is considered expensive. Various states such as Rajasthan, Karnataka, Tamil Nadu, Telangana, Andhra Pradesh, and Gujarat have performed well as compared to others in solar power generation in India.

India being a diverse country in terms of its geography, electricity regulations, and per capita income needs to adopt an integrated approach to address various concerns to achieve the enhanced generation of solar energy in India. It requires focusing on multiple stakeholders such as government agencies, distribution companies, consumers, private players, and manufacturers to address the issues in the low adoption of solar energy in India.

Notes

1 Power supply for agricultural purposes in Haryana and other states is priced at very low in the range of INR 0.25 per unit as compared to INR 6 to INR 10 for residential/commercial purposes.
2 AT&C losses indicate the losses incurring due to technical and commercial losses and non-realisation of the billed amounts of power supply.
3 Net-metering is a concept which is used to calculate the net balance of electricity used by a consumer from the power grid and the electricity supplied to the power grid through the solar system.
4 The state of Haryana has two DISCOMs (power distribution companies) – DHBVNL (Dakshin Haryana Bijli Vitran Nigam Limited) and UHBVNL (Uttar Haryana Bijli Vitran Nigam Limited).

References

DCOH (2011a). *District Census Handbook Sonipat*. Directorate of Census Operations Haryana. Retrieved from: https://cdn.s3waas.gov.in/s3d296c101daa88a51f6ca8cfc1ac79b50/uploads/2018/04/2018040930.pdf (Last accessed: 28.03.2021).

DCOH (2011b). *District Census Handbook Hisar*. Directorate of Census Operations Haryana. Retrieved from: https://cdn.s3waas.gov.in/s3e56954b4f6347e897f954495eab16a88/uploads/2018/03/2018030694.pdf (Last accessed: 28.03.2021).

DNRE (2016). *Haryana Solar Power Policy, 2016*. Government of Haryana New and Renewable Energy Department. Retrieved from: https://cdnbbsr.s3waas.gov.in/s3f80ff32e08a25270b5f252ce39522f72/uploads/2020/12/2020120946.pdf (Last accessed: 17.08.2020).

DNRE (2021). *Draft Haryana Solar Power Policy, 2021*. Government of Haryana New and Renewable Energy Department. Retrieved from: https://cdnbbsr.s3waas.gov.in/s3f80ff32e08a25270b5f252ce39522f72/uploads/2021/04/2021042434.pdf (Last accessed: 12.08.2021)

Dobrotkova, Z., Surana, K. & Audinet, P. (2018). The price of solar energy: Comparing competitive auctions for utility-scale solar PV in developing countries. *Energy Policy*, 118(2018): 133–148. https://doi.org/10.1016/j.enpol.2018.03.036

Dutt, D. (2020). Understanding the barriers to the diffusion of rooftop solar: A case study of Delhi (India). *Energy Policy*, 144(2020). https://doi.org/10.1016/j.enpol.2020.111674

Goel, M. (2016). Solar rooftop in India: Policies, challenges and outlook. *Green Energy and Environment*, 1(2): 129–137. https://doi.org/10.1016/j.gee.2016.08.003

HAREDA (2016). *Haryana Solar Power Policy, 2016*. Department of New and Renewable Energy. Retrieved from: https://hareda.gov.in/document/haryana-solar-power-policy-2016/ (Last accessed: 05.02.2022).

IEA (2021). *India Energy Outlook 2021*. International Energy Agency (IEA), Paris. February 2021. Retrieved from: https://www.iea.org/reports/india-energy-outlook-2021 (Last accessed: 15.12.2022).

IEA (2023). *Evolution of Solar PV Module Cost by Data Source, 1970–2020*. Retrieved from: www.iea.org/data-and-statistics/charts/evolution-of-solar-pv-module-cost-by-data-source-1970–2020 (Last accessed: 11.02.2023).

IREDA (2020). *Background*. Indian Renewable Energy Development Agency Limited. Retrieved from: www.ireda.in/background (Last accessed: 18.08.2020).

IRENA (2023). *Renewable Capacity Statistics 2023*. International Renewable Energy Agency, Abu Dhabi. Retrieved from: www.irena.org/Publications/2023/Mar/Renewable-capacity-statistics-2023 (Last accessed: 27.11.2023).

Karakaya, E. & Sriwannawit, P. (2015). Barriers to the adoption of photovoltaic systems: The state of the art. *Renewable and Sustainable Energy Reviews*, 49: 60–66. https://doi.org/10.1016/j.rser.2015.04.058

Mercom (2014). *Survey of India Consumers Perceptions on Renewable Energy*. Mercom Communications India. Retrieved from: https://mercomindia.com/wp-content/uploads/2018/01/Survey_MercomIndiaSurveyRenewables.pdf (Last accessed: 05.07.2020).

MNRE (2021a). *Solar Rooftop Calculator*. Retrieved from: https://solarrooftop.gov.in/rooftop_calculator (Last accessed: 18.08.2021).

MNRE (2021b). *Jawaharlal Nehru National Solar Mission – Towards Building Solar India*. Retrieved from: http://indiaenvironmentportal.org.in/files/mission-document-JNNSM.pdf (Last accessed: 17.08.2020).

MNRE (2021c). *Solar Schemas*. Retrieved from: https://mnre.gov.in/solar/schemes (Last accessed: 21.08.2020).

MNRE (2021d). *Annual Report 2020–2021*. Retrieved from: https://mnre.gov.in/img/documents/uploads/file_f-1618564141288.pdf (Last accessed: 24.01.2022).

MNRE (2021e). *Physical Progress until 31.10.2021*. Retrieved from: https://mnre.gov.in/the-ministry/physical-progress (Last accessed: 03.12.2021)

MNRE (2021f). *Solar RPO and REC Framework*. Retrieved from: https://mnre.gov.in/solar/rpo (Last accessed: 03.12.2021)

MNRE (2022). *Annual Report 2021–2022*. Retrieved from: https://mnre.gov.in/annual-reports-2021-22/ (Last accessed: 27.11.2023)

MNRE (2023a). *Programme/Scheme Wise Cumulative Physical Progress as on December 2022*. Retrieved from: https://mnre.gov.in/img/documents/uploads/file_s-1673341299172.pdf (Last accessed: 04.02.2023).

MNRE (2023b). *Solar Energy-Grid Connected*. Retrieved from: https://mnre.gov.in/solar/solar-ongrid (Last accessed: 04.02.2023).

NISE (2020). *About NISE*. National Institute of Solar Energy. Retrieved from: https://nise.res.in/about-us/about-nise/#page-content (Last accessed: 19.08.2020).

NPP (2022). *All India Installed Capacity of Power Stations*. National Power Portal. Retrieved from: https://npp.gov.in/public-reports/cea/monthly/installcap/2022/OCT/capacity1-2022-10.pdf (Last accessed: 10.12.2022)

OECD/UN (2018). *Production Transformation Policy Review of Chile: Reaping the Benefits of New Frontiers. OECD Development Pathways*. OECD Publishing, Paris. https://doi.org/10.1787/9789264288379-en

OEERE (2020). *Solar Radiation Basics. Office of Energy Efficiency and Renewable Energy*. Retrieved from: www.energy.gov/eere/solar/articles/solar-radiation-basics (Last accessed: 19.07.2020).

Parikh, A. (2020). *India's Solar Imports Nosedive in Q1 2020 Amid Lockdown Due to COVID-19*. Mercom India. Retrieved from: https://mercomindia.com/india-solar-imports-q1-2020/ (Last accessed: 27.07.2020).

PIB (2019). *MNRE Refutes Reports Expressing Doubt on India's Renewable Energy Target 'India would Achieve 1,75,000 MW Target on Time' Says Secretary MNRE*. Press Information Bureau Government of India. Retrieved from: https://pib.gov.in/PressReleseDetailm.aspx?PRID=1587655 (Last accessed: 20.11.2020).

PIB (2021). *India Has Achieved Its NDC Target with Total Non-Fossil Based Installed Energy Capacity of 157.32 GW which is 40.1% of the Total Installed Electricity Capacity*. Press Information Bureau Government of India. Retrieved from: www.pib.gov.in/PressReleasePage.aspx?PRID=1785808 (Last accessed: 25.01.2022).

Rathore, P.K.S., Chauhan, D.S. & Singh, R.P. (2019). Decentralized solar rooftop photovoltaic in India: On the path of sustainable energy security. *Renewable Energy*, 131. February 2019. https://doi.org/10.1016/j.renene.2018.07.049

REN21 (2022). *Renewables 2022 Global Status Report*. REN21 Secretariat, Paris. Retrieved from: www.ren21.net/wp-content/uploads/2019/05/GSR2022_Full_Report.pdf (Last accessed: 16.12.2022)

Sareen, S. (2021). Legitimating power: Solar energy rollout, sustainability metrics and transition politics. *Environment and Planning E: Nature and Space*, 5(3): 1014–1034. June 2021. https://doi.org/10.1177/25148486211024903

SECI (2020). *Introduction*. Solar Energy Corporation of India Limited. Retrieved from: www.seci.co.in/about/introduction (Last accessed: 18.08.2020).

Tyagi, V., Dwivedi, P. & Gupta, A. (2019). Solar rooftop PV systems – markets, policies and future potential. *International Research Journal of Engineering and Technology*, 6: 1686–1691. March 2019.

UDAY (2023). *AT & C LOSS (Aggregate Technical and Commercial Loss)*. Retrieved from: www.uday.gov.in/atc_india.php (Last accessed: 05.02.2023).

UNFCCC (2015). *India's Intended Nationally Determined Contributions: Working towards Climate Justice*. United Nations Framework Convention on Climate Change, Bonn. Retrieved from: https://www4.unfccc.int/sites/submissions/INDC/Published%20Documents/India/1/INDIA%20INDC%20TO%20UNFCCC.pdf (Last accessed: 08.07.2020).

Urpelainen, J. (2016). Energy poverty and perceptions of solar power in marginalized communities: Survey evidence from Uttar Pradesh, India. *Renewable Energy*, 85: 534–539. January 2016. https://doi.org/10.1016/j.renene.2015.07.001

World Bank Group (2020). *Global Solar Atlas 2.0*. Retrieved from: https://globalsolaratlas.indfo/download (Last accessed: 27.11.2023).

7 Wind Energy in India

Status Quo of Achievements, Challenges, and Future Prospects

Deepak Sangroya, Kishore Kumar Gangwani, and Sanjay Chaudhury

7.1 Introduction

Energy consumption plays an indispensable role in the development of an economy (Apergis & Payne, 2010; Bhattacharya et al., 2016). Since the early days of the Industrial Revolution, fossil fuels have been recognised as a highly potent, primary source of energy. As of date, fossil fuels constitute around four-fifth of global energy consumption (Irfan et al., 2019a).

Over-reliance on fossil fuels has led to concerns about environmental implications and energy security. The increasing usage of fossil energy sources contributes to harmful greenhouse gas (GHG) emissions causing global warming. Environmental pollution due to GHGs is triggering a plethora of problems for world economies, challenging governments, their regulatory bodies, and the industry to devise strategies that help alleviate the risks of environmental pollution and global warming. In fact, economies across the globe are facing the simultaneous challenge of boosting economic output while tackling environmental degradation. Fossil fuels provide non-renewable sources of energy, which will eventually fall out. There is sufficient evidence that supplying fossil fuels for consistent economic growth will be challenging. Fossil fuels may not be available in the next few decades and depletion may occur sooner than expected (Kumar et al., 2010).

Concerns about environmental pollution and the future availability of affordable fossil fuels have forced regulatory bodies and policymakers to focus on alternative, renewable sources of energy (Irfan et al., 2019b). The growing interest in renewable energies is evident from the latest IRENA Renewable Capacity Statistics, which show that more than 80 per cent of newly added electricity capacity globally is from renewable energies. These sources, which include wind energy, solar energy, biomass, and hydro power, are non-exhaustive in nature and significantly less harmful to the environment (Chaurasiya et al., 2019). Among renewable energy sources, solar and wind account for 91 per cent of new renewables installations (IRENA, 2021). Wind and solar energy sources are clear winners among these alternative energy sources, as wind and sunlight are available across the globe with certain variations (Nguyen et al., 2021). Their availability and harnessing

DOI: 10.4324/9781003281818-7

potential are dependent on geographical conditions, weather conditions, time of the day, and technological capacities (Ouammi et al., 2012; Liu, 2018).

As a sustainable and renewable energy source, cumulative installed wind capacity at the global level has shown a consistently upward trend over the past two decades. The global cumulative installed wind energy capacity grew from a meagre 24.3 GW in 2001 to impressive 742.7 GW in 2020 (GWEC, 2021). In the year 2020, more than 93 GW installed wind capacity was added, representing a substantial development of the global wind energy sector. A few countries show disproportionately larger shares of installed wind energy than others. These countries retain more than 80 per cent of worldwide wind energy and include China, the United States, Germany, India, Spain, the United Kingdom, Brazil, France, Canada, and Sweden (WWEA, 2022). Apart from China, India is the only Asian country to make it into the list of top ten countries with its cumulative wind energy installation.

7.2 Foundations of India's Sustainable Energy Future

Most nations are striving to achieve sustainable development, and energy is a critical component for achieving this objective. For India, which is now the most populous country in the world and home to one-third of the world's poor population, the need to sustain high economic growth rates is critical (Wang & Liu, 2021). The need for high growth rates is associated with proportionally higher energy consumption and demand. The gap in energy demand and supply, however, is prevalent. A massive population and corresponding economic growth have increased energy demand and intensified supply side challenges, which include an enormous burden on domestic thermal energy sources (Swain & Chandra, 2022).

India's natural energy resources are primarily coal-based, while the demand for petroleum and natural gas is met through imports (Mu Ramkumar et al., 2019). Based on current trends, it is expected that India's dependence on oil imports will increase to 90 per cent by the year 2040 (IEA, 2019). Oil price fluctuations are an additional challenge in the overall goal of achieving energy security. Further, India has low energy efficiency, which has resulted in higher GHG emissions. Globally, India is the third-largest contributor of GHGs (Zhu et al., 2018). There has been an increased pressure on countries to reduce their GHG emissions and to meet the larger share of the energy requirement through renewable energy sources.

India can solve the challenges of its current demand–supply gap, its dependence on oil imports, and its over-reliance on thermal energy, while contributing to achieve sustainable development both domestically and globally. With a shift to renewable energy sources, India responds to sustainable development goal (SDG) 7 – affordable and clean energy – and SDG 13 – climate action set by the United Nations. India is specially positioned to help achieve target 7.2, that is, "[b]y 2030, increase substantially the share of renewable energy in the global energy mix."

India enjoys an abundance of renewable energy sources and has vast developmental potential (Raghuwanshi & Arya, 2019). Annually, India receives sunlight in the range of 2,300–3,200 hours, which is equivalent to six million kWh of energy (Hairat & Ghosh, 2017). Similarly, the developmental potential for wind energy is more than 100 GW (Wang & Liu, 2021). At 26th Conference of the Parties in Glasgow in 2021, India committed to 500 GW from renewable sources of energy by 2030. As an intermediary target, the Government of India determined installing power capacity of 175 GW from renewable energies by 2022, of which 60 GW from wind energy.

7.3 Wind Energy in India

7.3.1 Wind Energy Potential

India has immense potential for wind energy. Wind energy potentials are site-specific. Hence an assessment is essential for the selection of project locations. The National Institute of Wind Energy has commissioned more than 900 wind-monitoring stations and generated maps for wind potential at different heights. The Ministry of New and Renewable Energy (MNRE) has estimated a gross wind power potential of 302 GW at 100 m and 696 GW at 120 m above ground level. Table 7.1 shows the wind power potential in different Indian states according to which Gujarat leads the list as the most wind-resource-endowed state, followed by Karnataka, Rajasthan, Maharashtra, Andhra Pradesh, Tamil Nadu, and Madhya Pradesh.

As the wind potential increases with height, it is desirable to utilise the 120-m height while increasing the size of wind turbines. Globally the size and height of installed wind turbines has increased over time (Serrano-González & Lacal-Arántegui, 2016). In that regard, India is like the rest of the world. However, with increase in height, the installation cost increases and logistical challenges for transportation of turbine blades and other components emerge

Table 7.1 State-wise potential wind capacity from highest to lowest.

Rank	*State*	*Wind potential at 100 m (GW)*	*Wind potential at 120 m (GW)*	*Total wind potential (GW)*
1	Gujarat	84.43	142.56	226.99
2	Karnataka	55.85	124.15	180.00
3	Rajasthan	18.77	127.75	146.52
4	Maharashtra	45.39	98.21	143.60
5	Andhra Pradesh	44.22	74.90	119.12
6	Tamil Nadu	33.79	68.75	102.54
7	Madhya Pradesh	10.48	15.40	25.88
	Others	9.28	43.78	53.06
	Total	302.25	695.5	997.75

Source: MNRE (2021)

(Cotrell et al., 2014). Once these challenges are addressed, wind powers at heights of 120 m and above can be utilised.

7.3.2 Installed Wind Power in India

India's wind energy-installed capacity has increased significantly in recent years. As of December 31, 2021, the total installed wind power capacity is 40.083 MW, which equals more than 40 per cent of total installed renewable energy capacity (MNRE, 2021; NPP, 2022). Seven Indian states contribute more than 99 per cent of installed wind power capacity in India (see Table 7.2). Tamil Nadu has the largest installed wind power capacity, followed by Gujarat, Maharashtra, Karnataka, Rajasthan, Andhra Pradesh, Madhya Pradesh, Telangana, and Kerala.

7.3.3 The Wind Energy Gap: Potential Versus Installed Capacity

As of December 2021, installed wind capacity was slightly above 40 GW. The biggest jump in cumulative wind power capacity amounted to 5.5 GW in 2015–2016. India is thus likely to fall short of its target of 60 GW installed wind energy capacity by 2022. However, growth in the sector has been remarkable thus far and, not least, because of the support by the Union Government. The Indian Government has supported various state governments to harness their wind energy potential through multiple programmes and projects (Mahesh & Jasmin, 2013). Table 7.2 presents latest available figures of the top seven Indian states in terms of utilising their wind potential.

Table 7.2 shows that India currently uses only 4 per cent of its wind energy potential with Tamil Nadu and Gujarat contributing the lion share. Surprisingly, the comparison of potential and installed wind capacity at state level

Table 7.2 State-wise ranking of wind potential utilisation.

Rank	*State*	*Total wind potential (GW) (Rank)*	*Installed wind power (GW) (Rank)*	*Utilisation of state/India total wind potential (%)*
1.	Madhya Pradesh	25.88 (7)	2.52 (7)	9.74/0.25
2.	Tamil Nadu	102.54 (6)	9.85 (1)	9.61/0.99
3.	Gujarat	226.99 (1)	8.95 (2)	3.94/0.90
4.	Maharashtra	143.60 (4)	5.01 (3)	3.49/0.50
5.	Andhra Pradesh	119.12 (5)	4.10 (6)	3.44/0.41
6.	Rajasthan	146.52 (3)	4.33 (5)	2.96/0.43
7.	Karnataka	180.00 (2)	5.04 (4)	2.80/0.51
	Others	53.06	0.19	0.37/0.02
	Total	997.75	39.99	4.01

Source: MNRE (2021)

reveals that the two listed tail light states in terms of wind potential – Madhya Pradesh and Tamil Nadu – perform much better than other states in utilising their wind energy potential. This indicates that they have created a more conducive environment and perform better in attracting investments for wind development projects. The following section highlights state-level developments that might explain the Table 7.2 ranking.

7.3.3.1 *Madhya Pradesh*

It is surprising that the landlocked state, Madhya Pradesh has been able to utilise a sizeable portion of its comparatively low wind potential. With almost 10 per cent of wind energy utilisation, Madhya Pradesh lands first in the state ranking. Even if not among the first movers in developing the wind energy sector, the state's government established a public–private partnership in 1994 to develop wind energy projects. Over the years, the government of Madhya Pradesh modified its wind energy policy to achieve ever-increasing wind energy installation targets in collaboration with private sector companies. Overall, the state's success would make an interesting case study to further our understanding of state internal governance mechanisms and their potential replicability.

7.3.3.2 *Tamil Nadu*

The coastal state of Tamil Nadu is a pioneer and the first to approach the threshold of 10 GW installed wind energy in the country. The state government realised the importance of alternative energy sources and set up the Tamil Nadu Energy Development Agency (TEDA) as early as 1984. Tamil Nadu's annual wind capacity additions have been regularly among the highest in the country. One of Asia's largest wind parks is the Muppandal wind farm group in the district of Kanyakumari. The achievements of the state can be attributed to continuous government effort in both assessing wind resource potential and opening that potential to private investors since early 1990. Tamil Nadu also stands out as an international manufacturing hub of wind turbines.

7.3.3.3 *Gujarat*

The state of Gujarat has a 1,600 km coastline with favourable wind speeds and the highest regional wind energy potential at its disposal. Like TEDA, the state of Gujarat set up the Gujarat Energy Development Agency in 1979 to enable a conducive environment for generation of sustainable energy. The state is one of the early movers in developing wind farms in India, starting operations as early as 1985–1986. Gujarat has witnessed a sharp increase in wind energy generation over the past few years, which can be attributed to the government's focus on tapping the abundance of wind energy. One effective government initiative was to provide land for wind power projects on lease basis, for example. At the "Vibrant Gujarat" summit in 2017, the Government of Gujarat presented several Memoranda of Understanding (MoUs) for projects worth INR 1 lakh crore (US$13.3 billion) in the renewable energy segment (TOI, 2017). During fiscal year 2020–2021, the increase in newly installed wind capacity was more than 1 GW (TOI, 2021).

7.3.3.4 Maharashtra

Maharashtra is one of the leading states ranking fourth in terms of total wind energy potential of 143.6 GW and third with more than 5 GW in installed wind power capacity. Despite its enormous potential, the utilisation of wind remains less than 4 per cent. Conscious efforts by the state government in making wind energy investments a viable option for developers are proven by an increasing number of wind energy developers and projects registered with the Maharashtra Energy Development Agency – a state nodal agency under the umbrella of the MNRE to undertake the development of renewable energy. Maharashtra has been rated as one of the best states to invest in wind energy due to investor-friendly policies. Almost all major wind turbine manufacturers have a presence in the state.

7.3.3.5 Andhra Pradesh

The southern state of Andhra Pradesh has done well in terms of wind energy utilisation. Its wind energy utilisation rate of 3.4 per cent is comparable to that of Maharashtra and is only slightly below the country-wide achievement rate. Different Independent Power Producers (IPPs) such as Mytrah Energy or the Greenko Group are active in the state. However, just as all other Indian states with high wind potential, Andhra Pradesh could do better in addressing the challenges inherent in the development of the wind energy projects.

7.3.3.6 Rajasthan

Since the initiation of Rajasthan's first 2-MW wind power project in 1999, the installed capacity has reached more than 4.3 GW as of October 2021. With the 1.6 GW Suzlon Energy wind park project in Jaisalmer district, the state features one of the biggest wind energy projects in India and entire Asia. Rajasthan has the country's third-highest wind energy power potential (146.52 GW) after its neighbouring states Gujarat and Karnataka. However, the state's installed capacity is 50 per cent less than that of Gujarat. With that Rajasthan ranks fifth in country-wide installed wind power capacity. The lack of policy support has stalled the growth of wind energy in the state. Over the past four years, only 45 MW wind power capacity have been added and tapping the state's wind energy potential remains complicated (TOI, 2020). Rajasthan is committed to promote all renewable energy technologies but seems to focus on solar installations – a sector in which it is outperforming other Indian states by far (MNRE, 2021).

7.3.3.7 Karnataka

The state of Karnataka ranks fourth with close to 5 GW of installed wind energy capacity. Karnataka is one of the very few states with a well-formulated policy for renewable energy generation. Currently, most of the energy demand in Karnataka is met through its thermal and hydro power plants situated in Raichur, Bellary, and Yeramarus. However, the state is progressive in moving away from thermal power production and increasing the contribution of renewable energies to meet its energy needs (NIE, 2021). In terms of its wind energy potential, Karnataka ranks second after Gujarat with a total of 180 GW. The state's wind utilisation rate remains below 3 per cent. Like the case of Rajasthan, Karnataka might favour solar energy over wind as showing in the state's high installed solar energy capacity.

Interestingly, none of the Indian states is close to harness even half of its total wind potential. Currently, only two Indian states are close to 10 per cent, whereas five Indian states are utilising only 2 to 4 per cent of their total wind potential. Increasing the utilisation of available wind potential remains a key task for state governments and regulatory bodies. Nevertheless, whatever has been achieved till date has been a remarkable journey. In the following section, we present the policy initiatives that led to the growth of India's wind energy sector.

7.4 Policy Initiatives for Wind Energy Uptake

In the past three decades, Indian wind energy sector has developed through regulatory intervention, central and state government monetary incentives, and private section investment. Encourage by government support, the private sector has invested heavily in wind power projects, particularly in the states of Tamil Nadu, Gujarat, and Rajasthan. Private sector investments have driven down costs and helped increase the country's wind energy capacity, making it the fourth-largest wind energy producer in the world. While government fiscal and monetary incentives make private sector investment in wind energy lucrative, there are also risks involved. One significant risk is associated with land acquisition and community engagement, which can result in delays and increased costs. Another risk is the regulatory and policy environment, which can be subject to change and may impact the financial viability of projects. Additionally, unpredictability of wind patterns is also a major challenge, which can affect the generation of electricity and revenue for the company. In spite of challenges such as land acquisition and transmission infrastructure development, private companies invested in wind energy and boosted local economies. Overall, private sector investors need to carefully make a trade-off between these risks and benefits (mentioned in the below section) to ensure the long-term success of their wind energy investments in India.

7.4.1 Accelerated Depreciation

One of the most successful policy instruments for the support of renewable energies is accelerated depreciation (AD) – a tax benefit which allows wind power project developers to claim depreciation equivalent to capital investment of the project, resulting in reduced tax obligation. When, in 1994, this scheme was introduced, it allowed AD at a rate of 100 per cent, which subsequently reduced to 80 per cent in 2002. This incentive was the most attractive for wind power producers and India witnessed an exponential growth in the installations of wind energy plants. However, these investments in wind energy were made primarily to claim tax benefits at the cost of both performance and quality of wind energy projects. As a result, installation increased but not generation. In 2012, the AD scheme was completely withdrawn.

7.4.2 Generation-Based Incentive

To improve generation from wind energy installations, the MNRE launched a scheme of providing a generation-based incentive (GBI) of INR 0.50 (US$0.007) for every unit of wind energy generated in 2010. In this scheme, wind energy projects were supported up to a maximum of INR 10 million per megawatt (US$133,000). This scheme was available only to companies or individuals such as IPPs who did not claim AD simultaneously. Like AD, this financial incentive attracted many wind energy investors, resulting in more than doubled annual wind energy installations between 2009 and 2011. However, the central government discontinued GBIs in 2017, citing financial reasons, that is huge losses to the exchequer.

7.4.3 Renewable Energy Certificates

In 2010, India's Central Electricity Regulatory Commission (CERC) introduced renewable energy certification. In this scheme, wind energy producers are getting one certificate for every 1 MWh of wind energy generated. Further, these renewable energy certificates (RECs) can be traded on the power exchanges, that is Power Exchange India Limited (PXIL) and Indian Energy Exchange Limited (IEX) within a predetermined limit of floor and forbearance price (Narula, 2013). Various obligated entities, like state utilities, direct access consumers and non-renewable captive generators can purchase RECs from IEX or PXIL to meet their renewable purchase obligation (RPO) (Chandel et al., 2016). However, the success of REC mechanism was hindered due to the lack of enforcement of RPOs (Kandpal & Dhingra, 2021). To strictly implement RPOs, target RPOs were defined and an RPO monitoring cell was established by the Ministry for New and Renewable Energies.

7.4.4 *Feed-In Tariff*

At the state level, wind power producers could sign a fixed-price, long-term power purchase agreement (PPA) with state governments (Gupta & Purohit, 2013). The State Electricity Regulatory Commission met industry expectations by taking the "cost-plus basis" approach for designing the wind electricity tariff. This approach considers several factors for determining the tariff, including capacity factor, lending cost, and capital cost. Along with the feed-in tariff (FIT), wind power producers could use tax benefits such as AD and GBI (Kar & Sharma, 2015). Since its introduction, the FIT became the most important policy instrument for attracting investments in wind energy projects because these long-term contracts were delivering a stable revenue stream. However, with the introduction of auctions in 2016–2017, state governments stopped offering FITs for new wind energy projects (Jethani, 2016).

7.4.5 *Reverse Auctions*

Wind energy sector in India has followed the global developments and replicated the global best practices for the overall development of the sector. Since 2017, auctions have been used to discover the lower tariffs of wind energy. Lower tariffs are due to the cheaper capital, increased competition, and large project size. Although the FIT has been an effective and successful support mechanism for the growth of wind energy, reverse auction (RA) has been introduced to lessen the subsidy provided by the government. RAs helped deploy wind energy in economically more efficient and transparent manner. Moreover, to establish a complete market-driven mechanism, the GBI scheme was discontinued and AD got reduced to 40 per cent when RAs were introduced. As mentioned earlier, the Solar Energy Corporation of India (SECI) conducted the first RA in the year 2017 and achieved a binding tariff of INR 3.46 per unit. Over a period of five years, a total of 12,853 MW of wind energy projects have been auctioned in which bids touched as low as INR 2.43/kWh (Kandpal & Dhingra, 2021).

7.5 Key Barriers to Wind Energy Adoption

India remains a lucrative fast-growing market for the wind energy industry. India needs to boost the wind energy generation to address the vast economic and social requirements (Rohankar et al., 2016). Compared to the global consumption, per capita consumption in India is just one-third of the global average (Bhide & Monroy, 2011). However, in the decade prior to the adoption of the SDGs and the Paris Agreement, that is between 2005 and 2015, this growth picked up and India's per capita electricity consumption increased to 1,010 kWh from mere 612 kWh, equalling a growth of more than 5 per cent per annum (Moallemi et al., 2014). This suggests fantastic opportunity for new generation capacity that needs to be developed over the coming decades. However, there are still various challenges which need to be addressed for a more smooth and speedy execution of wind energy projects. The most important challenges are discussed in the following subsections.

7.5.1 Financial Health of State Distribution Companies

In India, most of the generated wind energy is sold by electricity producers to state electricity distribution companies. However, these DISCOMs are in financial distress for many years. They are not able to pay their dues to energy producers on time. Delays between 3 and 18 months are common. Moreover, to avoid additional financial pressure, most DISCOMs are limiting their purchase of wind energy. As a result, financial conditions of wind energy producers are also deteriorating. Wind energy producers are dependent upon debt, burdened with high interest rates to pay the operational expenses of running wind energy turbines.

7.5.2 Renegotiation of PPAs

In the past few years, various state governments have started to renegotiate already signed PPAs with wind energy producers, despite clearly laid out policies and procedures by the MNRE. The state government of Andhra Pradesh, for example, has tried to renegotiate existing PPAs, citing malpractices in the process. They reduced the purchase of wind energy against what had been already agreed and committed to. This kind of renegotiation reduces the confidence of wind energy producers in state governments and makes it difficult to convince financiers to invest in wind energy projects. Also, Gujarat government cancelled and reissued various tenders citing higher tariffs in the bids. As a result, wind power producers had to cut down tariffs thereby reducing the financial viability of the projects.

7.5.3 Inadequate Energy Evacuation Infrastructure

Wind energy projects require infrastructure for power evacuation and transmission to the connection point of state DISCOMs. As the size of wind energy projects increases, the requirement for evacuation infrastructure also grows manifold. However, current evacuation infrastructure and transmission facilities are inadequate to carry wind energy from these mega size wind parks to DISCOMs. Many of even the best wind sites do not have evacuation facilities. Wind energy projects have a timeline of 18 months from the signing date of a PPA. However, the development of evacuation infrastructure takes around five years to become operational. In few other cases, wind energy project PPAs have been signed but evacuation work has not started even after the expiry of a deadline.

7.5.4 Land Issues

In India, land issues are regulated at state level and various agencies such as State Revenue Department, Gram Panchayat, and Forest Department are involved in approving land for wind energy projects. Due to the involvement of these various agencies, land procurement takes an exceedingly long time, sometimes even 18 months. The procedure of land procurement is cumbersome as wind energy producers must take various steps like conversion of land from agriculture use to wind energy use, taking approval from government agencies before they can get active on the project site. Land procurement is the most critical stage of wind energy project development and it involves various agencies resulting in unnecessary delay in the development of these projects. There is an urgent need for a comprehensive land procurement policy which addresses the demand of all key stakeholders in an integrated and thus more efficient way.

7.5.5 Tariff Caps

As a policy instrument, the Indian Government has decided to put an upper cap on wind energy projects during auctioning. Due to this upper ceiling, wind power producers cannot quote beyond the capped tariff rates while bidding for wind energy projects. Moreover, the tariff cap kept by SECI for wind energy projects is extremely low, which makes these projects financially unviable for wind power producers. These lower tariffs discourage wind power producers, and as a result, many auctions lapse unsubscribed. In fact, not a single auction between 2019 and 2020 was fully subscribed. Yet to impose an upper cap on the tariff is against the very principle of market auctioning. Auctions help the market determine the tariff rates based on demand and supply. SECI removed tariff caps in some of its tenders in March 2020, which resulted in reduced wind energy prices.

7.6 Strategies to Facilitate Wind Energy Adoption

After having presented major challenges, we now want to discuss strategies for increasing wind energy adoption in India. These strategies focus primarily on reducing hindrance and creating an investment-friendly environment for wind energy projects.

7.6.1 Grid Integration

For achieving higher wind energy installations, grid expansion is critical. Moreover, grid operation and management must become more efficient in the medium to long terms. The smooth operations of the grid can be achieved through proper coordination between state governments, DISCOMs, and wind power producers. Although the intermittent generation of wind energy poses challenges to grid integration, grid management technologies are proven solutions to handle the volatility. Accordingly, policies and orders issued by the CERC regarding evacuation arrangement and grid connectivity should be strictly followed by all authorities. We also suggest a central agency to test and certify all wind turbines accordingly as per advanced rules and standards.

7.6.2 Certainty of a Supportive Policy Framework

Although electricity is a state subject, a uniform policy considering site-specific wind potential should be created to increase investor confidence. Policies should continue for a longer period to bring stable growth into the wind energy market. By contrast, any policy changes should be done after consulting with all stakeholders in a transparent manner.

7.6.3 Land Availability

The process of land procurement for renewable energy projects, and especially wind energy projects, needs to be made simpler. The current approval format and related procedures need to be examined in detail and ways of streamlining them need to be identified for faster development of wind energy projects. Critical formalities such as conversion to non-agriculture use, acquisition of forest land, approval from wildlife sanctuaries, defence, and aviation could be significantly simplified. This would help wind energy developers to concentrate on the faster establishment and smoother operations of projects.

7.6.4 RPO Compliance and Regular Monitoring

The success of wind energy projects in the current policy framework is highly contingent upon the strict enforcement of RPOs by contractors. Subsequently, RPO enforcement will develop the power exchanges needed throughout the country. This binding of RPO compliance will immediately cause relief to many wind energy producers and will pave the way for more investments. Therefore, central agencies should ensure that state DISCOMs strictly fulfil RPOs in line with the requirements.

7.6.5 Offshore Wind Power Development

Meeting the growing need for renewable energies, the estimated offshore wind potential of India could lead to a much higher share in the overall electricity mix. With the deployment of offshore wind turbines, the proportionate contribution is expected to increase manifold. According to some preliminary estimates, India's total 7,600-km-long coastline has a potential of installing 140 GW offshore wind energy by 2050. However, regulatory authorities are required to formulate a policy framework and support programmes in a manner that attract offshore wind project investors. As offshore wind power development will have evacuation challenges like onshore wind energy, the grid would need further expansion along the Indian coastline.

7.7 Conclusion and Policy Implications

To overcome the current demand–supply gap in energy autonomy and to reduce GHG emissions, India has sharpened its focus on renewable energy generation. Wind power is one of the important options on the spectrum of renewable energy alternatives. India's success in realising an installed wind capacity of 40 GW even if not accurately on time can make a case study to be replicated in other regions of the world. Unlike solar power, wind power in India has been concentrated in a few regions thus far, with five states accounting for more than 80 per cent of India's installed wind power capacity. The southern state of Tamil Nadu is leading the ranks, with other states such as Gujarat, Maharashtra, Karnataka, and Rajasthan trying to catch up.

The growth in the Indian wind energy sector resonates well with policy and fiscal initiatives by the central and state governments, such as AD, GBI, REC, FIT, RA, and others. Despite these comprehensive policy initiatives, data suggest a wide variation across states. To maximise the full potential of wind energy and beyond, governments need to address challenges such as the financial viability of distribution companies; renegotiation of PPAs; inadequate infrastructure, including land as well as energy evacuation infrastructure; and tariff caps.

Although this chapter provides a comprehensive overview of current developments in India's wind energy sector, it has some limitations. First, our review suggests that social acceptance of wind energy projects is of critical importance for the growth of the sector. Like in many other countries, large public protests block the installation of wind energy projects in their localities (The Hindu, 2021). These protestors complain that birds and wildlife are disturbed by the rotor noise of the wind turbine. Therefore, public awareness and acceptance are crucial social factors that must be considered while developing wind energy plants. Second, even though local acceptance or opposition of wind energy installations by the local public can never be predetermined, strategies to successfully manage these stakeholders can be studied. In Europe, for example, key stakeholders have been interviewed to comprehend their views and generate empirical data on social acceptance (Enevoldsen & Sovacool, 2016). Moreover, these studies present actionable guidelines for policymakers and wind energy developers. However, hardly any such research exists for the Indian context. Hence, studying local stakeholder perspectives open avenues for future research.

References

Apergis, N. & Payne, J.E. (2010). Renewable energy consumption and economic growth: Evidence from a panel of OECD countries. *Energy Policy*, 38(1): 656–660.

Bhattacharya, M., Paramati, S.R., Ozturk, I. & Bhattacharya, S. (2016). The effect of renewable energy consumption on economic growth: Evidence from top 38 countries. *Applied Energy*, 162: 733–741.

Bhide, A. & Monroy, C.R. (2011). Energy poverty: A special focus on energy poverty in India and renewable energy technologies. *Renewable and Sustainable Energy Reviews*, 15(2): 1057–1066.

Chandel, S.S., Shrivastva, R., Sharma, V. & Ramasamy, P. (2016). Overview of the initiatives in renewable energy sector under the national action plan on climate change in India. *Renewable and Sustainable Energy Reviews*, 54, 866–873.

Chaurasiya, P.K., Warudkar, V. & Ahmed, S. (2019). Wind energy development and policy in India: A review. *Energy Strategy Reviews*, 24: 342–357.

Cotrell, J., Stehly, T., Johnson, J., Roberts, J.O., Parker, Z., Scott, G. & Heimiller, D. (2014). *Analysis of Transportation and Logistics Challenges Affecting the Deployment of Larger Wind Turbines: Summary of Results (No. NREL/TP-5000–61063).* National Renewable Energy Lab (NREL), Golden, CO, United States.

Enevoldsen, P. & Sovacool, B. (2016). Examining the social acceptance of wind energy: Practical guidelines for onshore wind project development in France. *Renewable and Sustainable Energy Reviews*, 53: 178–184. January 2016. https://doi.org/10.1016/j.rser.2015.08.041

Gupta, S.K. & Purohit, P. (2013). Renewable energy certificate mechanism in India: A preliminary assessment. *Renewable and Sustainable Energy Reviews*, 22: 380–392.

GWEC (2021). *Global Wind Energy Report 2021*. Global Wind Energy Council. Retrieved from: https://gwec.net/global-wind-report-2021/ (Last accessed: 07.12.2021)

Hairat, M.K. & Ghosh, S. (2017). 100 GW solar power in India by 2022–A critical review. *Renewable and Sustainable Energy Reviews*, 73: 1041–1050.

The Hindu (2021). Residents of Gujarat village protest against windmills being set up on forestland. *The Hindu*. 07.08.2021. Retrieved from: www.thehindu.com/news/national/other-states/residents-of-gujarat-village-protest against-windmills-being-set-up-on-forestland/article35781540.ece (Last accessed: 21.02.2021)

IEA (2019). *World Energy Outlook 2019*. International Energy Agency, Paris. Retrieved from: www.iea.org/reports/world-energy-outlook-2019 (Last accessed: 05.01.2021)

IRENA (2021). *Renewable Capacity Statistics 2021*. International Renewable Energy Agency (IRENA), Abu Dhabi.

Irfan, M., Zhao, Z.Y., Ahmad, M. & Mukeshimana, M.C. (2019a). Solar energy development in Pakistan: Barriers and policy recommendations. *Sustainability*, 11(4): 1206.

Irfan, M., Zhao, Z.Y., Ahmad, M. & Mukeshimana, M.C. (2019b). Critical factors influencing wind power industry: A diamond model-based study of India. *Energy Reports*, 5: 1222–1235.

Jethani, J.K. (2016). *Wind Power Development in India: An Overview*. Akshay Urja.

Kandpal, D. & Dhingra, T. (2021). Migrating to reverse Auction mechanisms in wind energy sector: Status and challenges. *Energy Policy*, 156: 112352.

Kar, S.K. & Sharma, A. (2015). Wind power developments in India. *Renewable and Sustainable Energy Reviews*, 48: 264–275.

Kumar, A., Kumar, K., Kaushik, N., Sharma, S. & Mishra, S. (2010). Renewable energy in India: Current status and future potentials. *Renewable and Sustainable Energy Reviews*, 14(8): 2434–2442.

Liu, Z. (2018). What is the future of solar energy? Economic and policy barriers. *Energy Sources, Part B: Economics, Planning, and Policy*, 13(3): 169–172.

Mahesh, A. & Jasmin, K.S. (2013). Role of renewable energy investment in India: An alternative to CO_2 mitigation. *Renewable and Sustainable Energy Reviews*, 26: 414–424.

MNRE (2021). *Physical Profess until 31.10.2021*. Ministry of New and Renewable Energy. Retrieved from: https://mnre.gov.in/the-ministry/physical-progress (Last accessed: 03.12.2021)

Moallemi, E.A., Ahamdi, A., Afrazeh, A. & Moghaddam, N.B. (2014). Understanding systemic analysis in the governance of sustainability transition in renewable energies: The case of fuel cell technology in Iran. *Renewable and Sustainable Energy Reviews*, 33: 305–315.

Mu Ramkumar, M., Santosh, M., Mathew, M.J. & Siddiqui, N.A. (2019). India at crossroads for energy. *Geoscience Frontiers,* 12(6): 100901. https://doi.org/10.1016/j.gsf.2019.10.006

Narula, K. (2013). Renewable energy certificates (RECs) in India – a performance analysis and future outlook. *Renewable and Sustainable Energy Reviews*, 27: 654–663.

Nguyen, V.C., Wang, C.T. & Hsieh, Y.J. (2021). Electrification of highway transportation with solar and wind energy. *Sustainability*, 13(10): 5456.

NIE (2021). Karnataka's renewable energy push. *New Indian Express*. 27.08.2021. Retrieved from: www.newindianexpress.com/opinions/editorials/2021/aug/27/karnatakas-renewable-energy-push-2350468.html (Last accessed: 15.09.2021)

NPP (2022). *All India Installed Capacity of Power Stations*. National Power Portal. Retrieved from: https://npp.gov.in/public-reports/cea/monthly/installcap/2021/DEC/capacity1-2021-12.pdf (Last accessed: 04.02.2022)

Ouammi, A., Ghigliotti, V., Robba, M., Mimet, A. & Sacile, R. (2012). A decision support system for the optimal exploitation of wind energy on regional scale. *Renewable Energy*, 37(1): 299–309.

Raghuwanshi, S.S. & Arya, R. (2019). Renewable energy potential in India and future agenda of research. *International Journal of Sustainable Engineering*, 12(5): 291–302.

Rohankar, N., Jain, A.K., Nangia, O.P. & Dwivedi, P. (2016). A study of existing solar power policy framework in India for viability of the solar projects perspective. *Renewable and Sustainable Energy Reviews*, 56: 510–518.

Serrano-González, J. & Lacal-Arántegui, R. (2016). Technological evolution of onshore wind turbines – a market-based analysis. *Wind Energy*, 19(12): 2171–2187.

Swain, A.K. & Chandra, R. (2022). *Band-Aid Solutions Won't Staunch India's Power Crisis*. The India Forum. 31.05.2022. Retrieved from: www.theindiaforum.in/article/band-aid-staunch-india-power-crisis (Last accessed: 15.08.2022)

TOI (2017). Vibrant Gujarat Summit: Rs. 1 lakh crore fuel to fire up renewables. *Times of India*. 08.01.2017. Retrieved from: https://timesofindia.indiatimes.com/city/ahmedabad/rs-1-lakh-cr-vgs-fuel-to-fire-up-renewable-energy/articleshow/56397050.cms (Last accessed: 15.09.2021)

TOI (2020). Wind power falls out of favour in Rajasthan, new capacity dwindles to 45MW in 4 years. *Times of India*. 31.12.2020. Retrieved from: https://timesofindia.indiatimes.com/city/jaipur/wind-power-falls-out-of-favour-in-raj-new-capacity-dwindles-to-45mw-in-4yrs/articleshow/80034592.cms (Last accessed: 15.09.2021)

TOI (2021). Wind power output in Gujarat nears record level. *Times of India*. 23.07.2021. Retrieved from: https://timesofindia.indiatimes.com/city/ahmedabad/wind-power-output-in-gujarat-nears-record-level/articleshow/84671518.cms (Last accessed: 15.09.2021)

Wang, Q. & Liu, Y. (2021). India's renewable energy: New insights from multi-regional input output and structural decomposition analysis. *Journal of Cleaner Production*, 283: 124230.

WWEA (2022). *World Market for Wind Power Saw Another Record Year in 2021: 97,3 Gigawatt of New Capacity Added*. World Wind Energy Association. 18.03.2022. Retrieved from: https://wwindea.org/world-market-for-wind-power-saw-another-record-year-in-2021–973-gigawatt-of-new-capacity-added/ (Last accessed: 17.11.2022)

Zhu, B., Su, B. & Li, Y. (2018). Input-output and structural decomposition analysis of India's carbon emissions and intensity, 2007/08–2013/14. *Applied Energy*, 230: 1545–1556.

8 The Power Distribution Sector

The Final Challenge to India's Green Transformation

Rishu Garg, E.V. Mallik, and G.V. Hanumanth Raju

8.1 Introduction

The Indian power sector is mainly organised in generation, transmission, and distribution sectors. Presently, each sub-sector is handled by a separate entity in various states. This, however, was not the case with electricity supply in the initial years. Over the years, the sector has undergone various changes with respect to the organisational structure and ownership. The existing distribution sector is mostly state-owned, with only 20 per cent of private companies operating of the total 71 distribution companies (DISCOMs). The following sections give an overview of how the Indian power sector and DISCOMs were institutionally shaped.

8.1.1 The Genesis of India's Distribution Companies

Back in the early nineteenth century, electricity was supplied primarily to urban areas and for railway traction. India's first power sector legislation, the Indian Electricity Act of 1887 provided a regulatory framework for supplying electricity for lighting and other purposes. The Act also called for attendants to look after the same and recognised the protection of persons and properties against electrical accidents. With the electrification of rural settlements, the Indian power sector initiated a number of reforms to safeguard consumer interests such as tariff-based energy consumption. The corresponding (re) formulation of many acts, policies, and schemes started to shape the institutional distinction of managing electricity through generation, transmission, and distribution sectors.

The first Act was revoked and substituted by the Indian Electricity Act of 1903, which laid general rules applicable to the whole country. However, the Act did not recognise the issue of dual control of the power sector by both local and Indian governments which led to a conflict of interest. This was resolved by The Indian Electricity Act of 1910, which also introduced the concept of issuing licences to companies for supplying bulk power to other small distributors. At the same time, the document enacted provisions to obtain distribution licences by the so-called distribution companies, that is

DOI: 10.4324/9781003281818-8

DISCOMs, which were mainly responsible for procurement of power from generators and retailing it to small consumers in a specified area. Additionally, the Act provided the legal framework for laying of wires and other related infrastructural works and safety concerns.

In independent India, another major Act was introduced. The Electricity Supply Act 1948 started to break up the central government's monopoly over the power sector and paved way for the creation of decentralised, vertically integrated State Electricity Boards (SEBs) and envisaged the creation of the Central Electricity Authority. SEBs hold the monopoly in the state's electricity business and were mainly responsible for infrastructure development and the coordination between generation, transmission, and distribution utilities to balance out gaps in electricity demand and supply. By the mid-1970s, SEBs poor operational performances led to deteriorating financial performances and losses in many states. Additionally, factors like high technical and financial losses, mismanagement, irrational tariffs, high subsidisation for agriculture and domestic consumers due to political appeasement, cross-subsidisations, poor generation capacity additions, and lack of transmission infrastructures aggravated the burden on DISCOMs. State governments have significant control over these SEBs with a single large entity covering generation, transmission, and distribution, which makes it difficult to catch the loss-making pockets. By the mid-1990s, this led state governments to disintegrate SEBs and create three independent entities for generation, transmission, and distribution. The rationale behind this process of "unbundling" was to isolate the three segments which would allow for focusing on the problems of each segment separately and for taking action appropriately. This promised to pave the way for more accountability and transparency in the system and intended to make DISCOMs economically viable. Moreover, SEB's restructuring led to the independent management of each sector's operations and finances.

However, the state governments were controlling SEBs directly such as through setting up irrational tariffs like high cross-subsidisation to industrial consumers and non-periodic revisions of state tariffs. The Electricity Regulatory Commission Act of 1998 introduced the Central Electricity Regulatory Commission and State Electricity Regulatory Commission (SERCs), two entities independent of central and state governments. SERCs were mainly responsible for determining tariffs for retail and bulk power supply in the state, generation charges, and transmission charges payable by DISCOMs to Generation Company (GENCOs) and Transmission Company (TRANSCOs).

The Electricity Act of 2003 was a major milestone in India's power sector reforms since it brought more openness and competition to electricity operations. This Act continued the unbundling of SEBs and aimed to complete the same by 2004. The Act further introduced competition through open access, the introduction of multi-year tariffs, distribution franchisees model, delicense of thermal generation with setting up renewable purchase obligation targets, emphasising the formation of regulatory bodies in the states, setting up an appellate tribunal for resolving disputes, and providing universal access to electricity.

8.1.2 DISCOMs and Their Ownership Structure

DISCOMs act as an interface between power producers and consumers. In the Indian Constitution, the power sector is classified as a concurrent subject where both central and state legislations can frame laws related to electricity. The obligation of supplying power to urban and rural consumers, however, lies with state governments.

The institutional structure of distribution companies plays a major role in the viability of the electricity market. Private participation in the distribution sector and horizontal unbundling of power markets have significantly impacted the operational and financial efficiency of DISCOMs. In most of the Indian states, DISCOMs are owned and operated majorly by state-owned entities, with few entities operated by private companies like TATA, ADANI, Reliance, Torrent, and some by state Power Departments. State-owned DISCOMs are responsible for power purchase agreements (PPAs) with generators and the retailing to the consumers of their area. DISCOMs normally procure electricity through short-term (3–12-month) contracts from power exchanges. They also allow consumers in their area to procure power through open access from energy providers other than the DISCOMs. Here, DISCOMs own grid assets and incur capital expenditure to supply electricity to their customers. They administer and maintain the billing of their consumers and collect the revenue by applying predetermined tariff rates as approved by SERCs.

There are two models for private participation in electricity distribution, the Distribution Franchisee Model and the Distribution Licences Model.

1. **Distribution Franchisee Model**
   In this model, a private party has no ownership rights on distribution assets and predominately manages billing and revenue collection in the area. For example, Torrent Power, Bhiwandi, Maharastra.
2. **Distribution Licensee Model**
   In this model, the private party holds equity and is a complete/partial owner of the distribution assets. The licensee here distributes electricity along with billing and collection of revenue. For example, TATA Power, New Delhi.

By the end of fiscal year 2022, India comprised 71 power distribution utilities, of which 45 were state-owned DISCOMs, 14 private DISCOMs, and 12 power departments. The distribution utilities are rated Y-o-Y by the Power Finance Corporation (PFC) on various parameters such as aggregate technical and commercial (AT&C) losses, revenue, average cost of supply (ACS)–average revenue realisation (ARR) gap, and debt. The latest PFC report rated

Table 8.1 State DISCOM structure.

Vertically bundled power departments	*Bundled with GENCO and DISCOM*	*Unbundled with one DISCOM*	*Unbundled with multiple DISCOMs*	
Arunachal Pradesh	Himachal Pradesh	Tripura	Andhra Pradesh	Maharashtra
Goa	Kerala	Chhattisgarh	Bihar	Odisha
Jammu and Kashmir	Manipur	Jharkhand	Delhi	Rajasthan
Mizoram	Punjab	Meghalaya	Gujarat	Uttar Pradesh
Nagaland	Tamil Nadu	Uttarakhand	Haryana	West Bengal
Pondicherry	Tripura		Karnataka	
Sikkim			Madhya Pradesh	

57 utilities (45 state-owned and 12 private) and 11 power departments. While 67 per cent of private utilities were rated A and A+, only 24 per cent of state utilities could achieve that rating. Similarly, only 27 per cent of power departments were given A ratings (MoP, 2023).

The operations of the distribution sector are complex due to different modes of ownership and a wide variation in DISCOM performance. Over the years, specific reforms were initiated to increase uniformity and improve performance, which we will present in the next section.

8.2 Reform Schemes of the Distribution Sector

Many of India's DISCOMs have generated high AT&C losses, which has led to their financial insolvency and has made them one of the weakest links in the Indian power system. Over the years, central and state governments have initiated various reforms to improve financial and operational performance to put the distribution sector on the right track. Distribution reforms cover various aspects from system improvements, AT&C loss reduction, theft prevention, consumer focus, commercialisation, and IT adoption. The following schemes were launched to address the issue of AT&C losses in the distribution sector. Several governments also implemented plans and bailout packages to address the issue of operational and financial inefficiencies in the sector.

8.2.1 Technical Reforms

In FY2001, the Ministry of Power (MoP) launched the Accelerated Power Development Programme (APDP) to improve the reliability of power supply at the distribution end. During the Tenth Five Year Plan (2002–2007), the APDP was relaunched as the Accelerated Power Development and Reforms Programme (APDRP) in 2002 with "Investment" and "Incentive" components introduced into the scheme, aiming to achieve commercial viability by minimising AT&C losses to 15 per cent.

To overcome the limitations of APDRP with respect to calculation of AT&C losses, significant deficiencies in maintenance of records, the absence of proper guidelines to DISCOM staff and several instances of incorrect reporting of AT&C losses by states, the Restructured Accelerated Power Development and Reforms Programme was launched in 2008 with revised terms and conditions and primarily implementing IT applications in the distribution sector. The programme covered areas of urban-towns and cities of population more than 30,000. The scheme also covered specific rural areas with comparatively higher load density and formulated the instruction of bifurcating agricultural load from domestic and industrial feeders.

On November 20, 2014, the Integrated Power Development Scheme (IPDS) received MoP's approval to improve the sub-transmission and distribution network, as well as metering at all levels of the metropolitan areas. The new IPDS scheme subsumed the previous Restructured Accelerated Power Development and Reforms Programme and almost a decade later is still work in progress.

8.2.2 Financial Reforms

The distribution sector was saddled with huge debts and financial losses. The financial initiative – a one-time settlement scheme – was introduced in 2001 to clear outstanding dues to the tune of INR 41,000 crores (roughly US$5 billion[1]) of state DISCOMs. Later in 2012, the Government of India approved another scheme named Financial Restructuring Plan to improve the financial health of DISCOMs, since total accumulated debt had reached INR 1.9 lakh crores (more than US$23 billion). Under Financial Restructuring Plan, state governments had to acquire 50 per cent of the outstanding short-term liabilities of DISCOMs. This was to be converted into bonds to be issued by DISCOMs to participating lenders duly backed by the state governments' guarantee. The remaining 50 per cent were to be restructured by rescheduling loans and providing moratorium on principal.

Even then the Indian power distribution sector continued to suffer losses and by the end of March 2015 it had an outstanding debt of INR 4.3 lakh crores (more than US$52 billion). The total accumulated losses in the sector reached INR 3.8 lakh crores (more than US$46 billion) in FY2015, an increase of 25 per cent over the accumulated losses incurred up to FY2014. In response, the Ujwal DISCOM Assurance Yojana was launched in November 2015 for the financial turnaround of DISCOMs. The scheme was developed and implemented to help DISCOMs deal with their financial and operational crisis.

Seven years later, since Ujwal DISCOM Assurance Yojana could not achieve the desired results, the Revamped Distribution Sector Scheme was launched in 2022. The reform-based and result-linked financial assistance was created to strengthen DISCOMs' supply infrastructure. The reform scheme has an outlay of INR 3,03,758 crore (approximately US$37 billion) over a five-year period (FY2022 to FY2026) and attempts to accomplish the following goals:

- By FY2025, AT&C losses will be reduced to 12–15 per cent at pan-India level.
- The gap between cost of supply and average realisable revenue (ACS–ARR gap) is to be eliminated by FY2025.
- With a financially viable and operationally effective distribution sector, customers will receive better quality that is reliability and affordability of power supply.
- The DISCOMs will be assessed using a "result evaluation matrix." Pre-qualifying conditions must be strictly met in order for funds to be released under the scheme based on the performance in the evaluation matrix.

Various reforms have been launched to improve DISCOMs' performance. The latest PFC report provides slight hope regarding AT&C or operational losses reducing to 16.5 per cent in FY2022 from 21.5 per cent in FY2021, and the gap between ACS and ARR improving from 70 paise per unit in FY2020 to 40 paise per unit in FY2022. However, the sector continues to bleed with DISCOMs having a liquidity gap of roughly 3.03 lakh crore (US$37.1 billion) and debt of roughly 6.2 lakh crore (US$75.8 billion). Unless there are stringent administrative practices deployed along with penalty and incentive mechanism for performances, we may have to witness another round of reforms in another five years.

8.3 DISCOMs Functioning and Operations

DISCOMs or the SEBs are responsible for distributing electricity to end-consumers. They are also responsible for maintaining the distribution system, which includes transformers, switchgears, poles, distribution lines, and other equipment. In addition, the billing and collection of revenue for electricity supply comes under their purview. The operations of distribution companies can be broadly classified into technical and financial operations. The following sections take a closer look at both these parts.

8.3.1 Technical Operations

The electricity generated at generation stations, that is power plants with conventional and/or non-conventional sources, is transferred to substations near load centres. That is where the operations of distribution utilities start as illustrated in Figure 8.1. End-consumers cannot use the high-voltage electricity as received at substations. It must be transformed to a lower voltage level that is appropriate for consumers. Hence a step-down power transformer

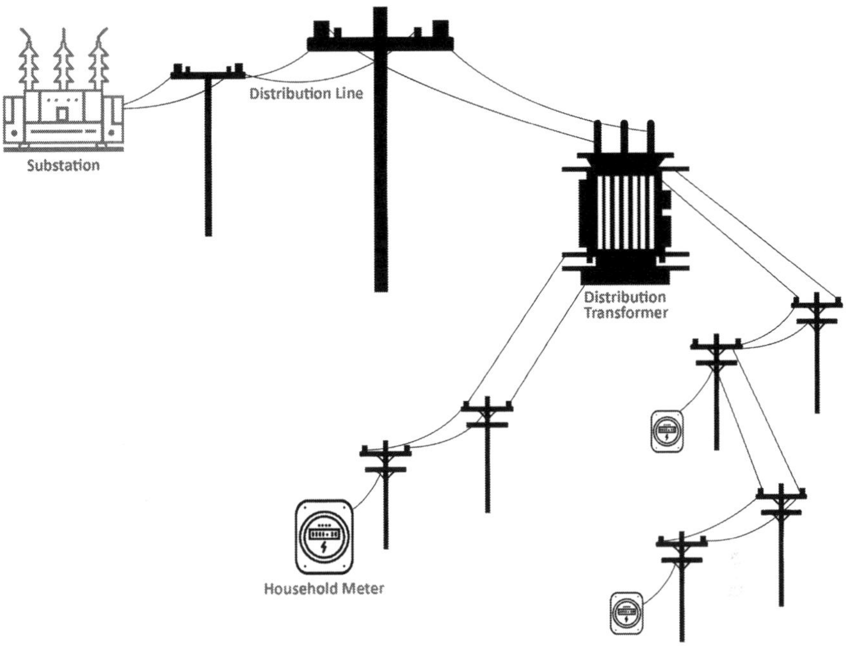

Figure 8.1 Typical distribution network.

is used to transform the voltage to 11 kV. The lines emanating from the substation are called "Distribution Feeders" or primary lines that carry electricity across load centres. The voltage level is further brought down to 440 or 230 volts at which most consumer appliances operate. This is done by "distribution transformers" (DTs) located near consumer areas. From there, electric poles carry the lines from the DTs to the consumer. A service line runs through either underground cables or pole-based overhead service wires to connect with the metre box at the building. The electrostatic meter at the consumer premises records consumption.

Distribution companies ensure to provide reliable electricity supply to end-consumers. As a part of their maintenance activity, they regularly inspect various equipments in the network to ensure smooth working conditions. They also carry out preventive maintenance drives in order to provide uninterrupted supply. In order to cater to growing demand, they plan the expansion of the network and upgradations of transformers and distribution lines.

8.3.2 Financial Operations

Financial operations play a critical role in the long-term sustainability of DISCOMs. The ACS and ARR are the two central indicators vital to DISCOM finances.

ACS is the average cost incurred by DISCOM to supply one unit of electricity to its consumer. A number of cost components are involved in order to arrive at ACS, which are detailed as follows:

(1) *Power purchase cost*: The cost incurred by DISCOMs towards power purchase from generating companies through PPAs or from the energy markets like Indian energy markets Ltd (IEX).
(2) *Transmission cost:* This includes the cost payable by DISCOMs to the state-owned transmission companies and also PGCIL for utilising its network for transmission of power from various sources. It also includes charges payable for the state load dispatch centres for operation and monitoring.
(3) *Operation and maintenance cost:* This includes expenditure made by DISCOMs towards the distribution cost, which involves the repair and maintenance of network assets like distribution lines, transformers, and other associated equipment.
(4) *Management and administration:* This includes the cost of maintaining the operations and administrative staff of the DISCOM and also cost incurred on general expenses.
(5) *Capital expenditure:* The investment made by DISCOMs on infrastructure such as electricity poles, transmission lines, and DTs.
(6) *Interest on capital borrowing and finance charges:* The DISCOMs incur interest for the loans, working capital, and for delayed payments to generators towards the power purchase bills. This also includes the other financial charges payable by DISCOMs.
(7) *Taxes and levies:* This covers various taxes and levies paid by DISCOMs to the government.

All these components taken together make the overall cost incurred to supply power to consumers with the unit cost calculated on the basis of total energy sold.

The ARR comprises revenue realised by DISCOMs through the following three streams:

(1) *Revenue from tariff:* This includes revenue collection from electricity consumers for energy sales and the settlement of accounts.
(2) *Subsidy:* The revenue received by DISCOMs from the state governments for providing free or subsidised power to a set of consumers.
(3) *Other incomes:* The revenue recovered from DISCOM assets and other sources.

With this, DISCOMs arrive at the average cost incurred to supply power to consumers and the revenue realised through the energy sale to consumers. The gap between the ACS and the ARR determines profit or loss, which in turn indicates the financial viability of DISCOM.

8.4 Challenges Faced by DISCOMs

Over decades, the Indian power distribution sector has faced myriad challenges of operational, financial and sociopolitical natures. There have been a few specific issues compromising the performance of the sector which will continue to haunt DISCOMs in coming years.

8.4.1 Operational Challenges

Technical losses in the distribution network pose a major hurdle in the sector. These high losses in the system depict the inefficiency in DISCOM operations. Key reasons for these losses are as follows:

Line losses: DISCOMs cater load in both urban and rural areas. When it comes to rural areas, the load is located far away from the substations. This results in increased line length of both primary lines (11 kV) and secondary lines (415 kV). The increase in secondary line length due to DTs placed far away from load centres can result in voltage issues and higher losses due to high resistance from lines stretched over longer distance.

Transformer loading: DISCOMs do not monitor the loading of transformers which normally results in overloading and increased losses and transformer failure. In addition, unbalanced loading in three phases can add on to losses in the system.

Poor maintenance: Maintenance of the distribution network is one of the common issues raised across the country. Some of the common issues are discussed as follows.

We tend to see plants and bushes growing near transformers, poles, and overhead electrical lines as illustrated in the photos below. This might lead to electrocution, injuries, and power outages when plants/bushes come in contact with the electric lines, which pose serious concerns on safety.

Poor maintenance of the distribution assets has raised concerns on the efficacy of DISCOMs' operations. The lack of maintenance of DTs would often lead to outages. Also, inadequate maintenance of DT centres has allowed for animals such as rats and birds to spread, which makes it more difficult for DISCOM workforce to access the location.

Inaccurate data management: Data management has been a huge task for DISCOMs. With the move towards technology integration, metering of feeders, DTs, and consumers have raised concerns over the effective monitoring and management of data. There are a few issues commonly noted in the sector.

- There is a move towards digitisation. Most of the reporting process is under manual interventions, however, which often leads to manipulations and calculation errors.

Figure 8.2 DT meter burnt-out box.

Figure 8.3 Poor condition of meter box.

Figure 8.4 Broken pole.

Figure 8.5 Poles covered with bushes – example 1.

Figure 8.6 Poles covered with bushes – example 2.

Figure 8.7 Inaccesible DT location.

Note: Photos illustrating maintenance issues in distribution network assets

Photo credit: E.V. Mallik

- Most of the reports generated often contain a wide range of discrepancies, as there are no secondary checks or third-party validations.
- Even though electricity bills are digitised, the billing happens in an old-fashioned way through the spot billing process. This is an inaccurate measurement of data process.

Lack of skilled labour and training: While the sector has been evolving with new technologies, DISCOMs have often failed to support their personnel through proper capacity-building programmes. The lack of skilled workforce has turned even technological advancements into disastrous failure. One such example is the GIS mapping of distribution assets under the Restructured Accelerated Power Development and Reforms Programme, which aimed at digitising all distribution assets in towns by specifically locating through geo-coordination and linkage of associated data. Due to the lack of skilled workforce, DISCOMs were unable to benefit because the data were not updated regularly. Moreover, DISCOMs were unable to extend the programme to semi-urban and rural areas.

8.4.2 Financial Challenges

Financial challenges of India's DISCOMs are of different kind, some of which are more recent like the transition to higher shares of renewable energies (REs); others have emerged over the years and decades as described below.

The commercial loss component involves two major factors. One is the inaccurate billing, which fails to capture the exact consumption of the consumer; the other one is unaccounted energy which normally happens due to theft.

- Reasons for inaccurate billing are due to faulty meter reading, unbilled installation due to default meter, or lack of meter reading.
- Loss of energy due to direct hooking and tampering of meters would result in commercial losses as there is no system to record the energy.

While DISCOMs supply as per needs and bill the installation costs, revenue collection poses a huge challenge. Even with various options and modes of payment available, consumers delay or default their payments, which reflects in increased AT&C losses. In fact, default payments have accumulated huge financial losses for DISCOMs over decades.

DISCOMs incur huge amounts towards development of infrastructure. However, even with upgrading the systems, losses in the system continue to hamper DISCOMs. Capital expenditures are not reflecting any breakthrough in addressing financial challenges nor do they add value in curtailing the issues of the sector.

One of the primary issues of DISCOMs is the increasing cost of supply. DISCOMs across India have signed long-term PPAs with thermal power plant operators and generally pay both fixed (must pay even if not used) and variable costs. With the RE share increasing in the system under a must-run status, DISCOMs are forced to back down their purchase from these thermal plants, which still contributes to the increase in cost of supply.

8.4.3 Sociopolitical Challenges

Other critical issues pertaining to the malaise of Indian DISCOMs concern the preferential treatment of India's agricultural sector and touch on questions of social justice.

In India, there is a provision for free power supply to the agricultural sector. The cost incurred to DISCOMs for supplying power to these consumers is borne by respective state governments, which release grants in the form of subsidies. However, the release of grants by the governments is delayed, which widens the financial gap and increases the accumulation of losses year on year. Moreover, we argue that the political decision on free electricity supply is the root cause that hinders the optimal use of water sources and the introduction of best practices in the irrigation sector.

Generally, DISCOMs carry out the tariff revision process where they arrive at applicable cost to consumers of various categories based on projections made in terms of energy and cost. These will be presented to the respective SERC. But due to the political and social influences of specific consumer categories, the approved tariff does not reflect the actual cost and tends to be on the lower side for set of categories (domestic agriculture and other institutions). This has been a reason of concern to DISCOMs, as it prevents them to charge the consumers based on the actual cost incurred.

The tariff for commercial and industrial categories also includes cross-subsidies. These are collected from commercial and industrial consumers to compensate the gap created by subsidising domestic categories, that is primarily consumers with income under the poverty line, the agricultural sector, and other benefitting categories. Subject to high tariffs, commercial and industrial consumers move towards open access, which leads to DISCOMs losing their high-paying consumer base.

8.5 Key Role of DISCOMs in the Energy Transition

In this section, we provide the readers with a glimpse on how the sector would evolve with technological innovations and by adopting smart operations. Here, we discuss a few levers that will drive the sector in the future.

8.5.1 Emerging Technologies

8.5.1.1 Smart Grids

The future of the power sector is in smart grids. With technological advancements, smart grids seem to be a reliable and more efficient solution to serving consumer needs. Smart grids promise to use resources more efficiently, to reduce losses in the system by bringing down the cost of supply, and to adopt smart operations. In addition, there is a wide range of opportunities as the sector is set to evolve with advancement in the following key areas.

8.5.1.2 Electric Vehicles

India's transport sector contributes about 10 per cent of total national greenhouse gas (GHG) emissions. The bulk of that (about 87%) is generated in road transport. Electric vehicles (EVs) can significantly reduce emissions in the transport sector. The Indian Government therefore intends to grow EV sales that would capture a market share of 30 per cent of private cars, 70 per cent of commercial vehicles, and 80 per cent of two- and three-wheelers by 2030. Both central and state governments initiated various schemes to promote the uptake of EVs as detailed in Chapter 11 ("Business Model Innovations for Electric Vehicle Adoption in India: An Ecosystem Perspective").

Clearly, the EV integration will have huge impact on the distribution sector with electricity demand steadily increasing. At this early stage of the market, energy demand from charging stations is unpredictable. While consumers can charge their vehicles at various locations and for different periods of time depending on their needs, the skewed nature of power demand and more widespread use of EVs may result in unintended peaks in power consumption. This could result in power quality problems such as increased flow in power cables, transformer overload, voltage drop, and voltage imbalance. DISCOMs that are responsible for developing the requisite charging infrastructure to support large-scale EV penetration must estimate the number of EVs that are likely to be in use and their estimated impact on the grid. The total electricity demand to power EVs is projected to increase to almost 640 TWh by 2030. The stability of the grid and reliability of power could be compromised if all EVs are connected to the grid at the same time, causing grid overload. This can get worse at times when the grid is already experiencing peak demand, and in locations where the network is already experiencing congestion. DISCOMs should adopt smart EV charging to avoid, shift, or delay peak loading conditions. If the charging is managed properly, EVs could effectively contribute and assist DISCOMs in managing the distribution network in an efficient and reliable manner.

8.5.1.3 Rooftop Photovoltaic

Government of India has set an ambitious target of 100 GW solar photovoltaics by 2022. The 100 GW target was further divided into 60 GW of utility scale solar installations and 40 GW of distributed rooftop photovoltaic (RTPV) installations. While the utility solar segment saw relatively rapid growth with 40.1 GW by 2022, RTPV could reach only 2.5 GW in the same span of time. The slower uptake of RTPV is attributed to reasons such as affordability (high upfront costs), lack of standardised products, lack of awareness among consumers, and issues of financing. However, RTPV systems have substantial benefits such as providing peaking power supply, reducing Transmission & Distribution (T&D) losses, improving tail end voltages, and creating local jobs. With RTPV being a major component of India's energy transition, DISCOMs should play a more prominent role in the proliferation of RTPV, particularly through client acquisition or demand aggregation.

8.5.1.4 Peer-to-Peer Trading Platforms

One of the ways to increase the uptake of rooftop solar is by taking a distributed approach where both small producers and consumers would be allowed to be on an electricity trading platform. This would enable prosumers to obtain economic benefits for the electricity generated near to load centres. The neighbourhood consumer could participate in electricity trading by purchasing the electricity at "community tariff," which could be lower than the tariff offered by DISCOMs.

In the transition to cleaner sources of energy, a number of peer-to-peer trading pilots and studies have been carried out in various countries. These studies cover business models, the role of the generator, and micro-grid-based community trading. A New York-based start-up, for example, came up with a blockchain-based community trading and micro-grid pilot study (Mok, 2016). Similarly, British company Piclo takes a collaborative approach where business consumers buy locally generated renewable power. Even in India, such pilot studies have been initiated in Uttar Pradesh and Delhi (Lama, 2019; Largue, 2019). However, decentralised electricity trading is an emerging concept, the advantages, disadvantages and socio-economic fitting of which still need to be better understood.

Different states have come up with solar electricity policies and tariff mechanisms for prosumers. Tariffs and cross-subsidy schemes vary from state to state for different categories of consumers. With different models in place, an evaluation would generate valuable insights into favourable trading mechanisms.

Basically, a decentralised energy-trading platform could be adopted for increased uptake of solar rooftop, and as an additional source of income for prosumers. A tariff structure that provides different rates at different times of the day is already in place for industrial and commercial consumers. With the roll-out of smart meters, the time-of-day tariffs could be implemented for domestic consumers as well. Once time-of-day tariffs are implemented at all levels, consumers would be on the lookout for better options to buy or sell power, which would help them to earn or save money through electricity trading.

For prosumers it can provide better prices for their energy generated at decentralised level near to load centre. For consumers, it would help to buy bulk energy more cost effectively compared to what the DISCOM tariff structure has to offer. DISCOMs, in turn, would get maximum energy at load centres, which would minimise the transmission and distribution losses. This could also result in better demand-side management (DSM) with the charges obtained from both prosumers and consumers for grid handling and service costs. This would reshape the electricity market by empowering consumers to participate in electricity trading.

8.5.1.5 Demand-Side Management

India's clean energy transition is influenced by the influx of intermittent and variable RE, mainly solar and wind. DSM will thus play a tremendous role in India's clean energy transition by reducing the demand and shifting consumption to times when REs can be effectively utilised. Further, DSM can help in improving end-use energy efficiency by enabling meeting part of the demand without having to increase supply. Proper DSM can help DISCOMs avoid purchasing costly electricity. Energy Efficiency, that is reducing overall energy consumption, Peak Load Management, that is reducing peak demand consistently over a season, and Demand Response, that is reducing peak demand for short periods of time for a few days during the year, are three major components of DSM. DSM should be encouraged as a DISCOM programme to fine-tune consumer's energy consumption pattern according to the DISCOMs' energy demand and distribution capacity without compromising consumers comfort. There is a need to ascertain specific targets under DSM with penalty on non-compliance for India's smooth transition to a cleaner energy future.

8.5.1.6 Behind-the-Meter Storage

A behind-the-meter storage (BMS) system, by definition, is a battery storage system connected to the distribution system placed behind the meter at the consumer premises. For decades, consumers are using uninterrupted power supply systems in order to obtain continuous power supply during outages. In the era of technological advancements, the concept of BMS is catching consumer interest as it holds multiple benefits:

- Better power quality and reliability of supply to consumers.
- Brings down electricity bills as the battery can release electricity during high tariff periods and get charged during lower tariff periods.
- Grid-friendly, as it has dual capability to respond to peaking needs (through discharge of energy) and shift to charging mode during off-peak time.
- Provides financial benefit to consumers through its grid-balancing services.

In order to reap the benefits, there is a need for development of suitable infrastructure. BMS requires advanced metering infrastructure in place along with the efficient monitoring and secure platforms. Communication systems would be at most priority as these systems needs accurate signals during the dispatch. The key factor to boost the uptake of these systems would be the time-of-day tariff setting, which would nudge consumers towards BMS. In addition, DISCOMs need to compensate consumers by providing suitable financial support through subsidies, availing loans, and simplifying the entire process.

BMS would be vital in shaping the energy transition. To identify and test the benefits of BMS, DISCOMs need to be proactive in initiating pilot projects that would allow them to better assess its diverse usage. Only then, they can and should plan ahead for BMS systemic mass integration.

8.6 Way Forward for DISCOMs

India has committed to achieve net-zero GHG emissions by 2070. One of the milestones of this bigger goal is to achieve 40 per cent of the energy from non-fossil fuel energy capacity by 2030, which would translate to approximately 500 GW of RE by 2030. Such tremendous integration of RE requires robust and flexible systems at both transmission and distribution scales. As the consumer facing entity, the distribution sector plays a major role for India's clean energy transition. However, with DISCOMs struggling massive financial and operational inefficiencies, no reforms or bailout schemes can usher India into an RE-dominant future before resolving issues at the distribution level. With further influx of evolving frontiers such as EVs and RTPV, DISCOMs must handle the three Ds: decarbonisation, decentralisation, and digitalisation. This requires a paradigm shift in their operations, planning, demand forecasting, and energy utilisation in the grid.

Note

1 Applied exchange rate 1US$ = INR 81.7602.

References

Lama, S. (2019). *Blockchain P2P Solar Energy Trading Goes Live in India*. 02.12.2019. Retrieved from: https://beincrypto.com/blockchain-p2p-solar-energy-trading-goes-live-in-india/ (Last accessed: 28.04.2023)

Largue, P. (2019). *Indian Electricity Distributor to Trial P2P Solar Trading*. Power Engineering International. 16.11.2019. Retrieved from: https://www.powerengineeringint.com/digitalization/indian-electricity-distributor-to-trial-p2p-solar-trading/ (Last accessed: 28.04.2023)

Mok, K. (2016). *Brooklyn Microgrid: A Blockchain-based Platform for Locally Traded Electricity*. Retrieved from: https://thenewstack.io/brooklyn-microgrid-blockchain-based-platform-locally-traded-electricity/ (Last accessed: 21.04.2023)

MoP (2023). *Urja Drishti – DISCOM Rating for Integrated Solvency, Health and Transparency Improvement. An Initiative by the Ministry of Power to Improve Transparency into India's Distribution Sector and Strengthen the Country's Discoms*. Retrieved from: https://urjadrishti.com/ (Last accessed: 07.05.2023)

9 Energy Storage

Enabling a Greener Grid

Rahul Walawalkar and Debmalya Sen

9.1 Introduction

"The Only Constant in Life Is Change" – a well-known phrase by Greek philosopher Heraclitus has become the mode of working in today's world. Innovations and disruptions have made many sectors rethink the way they operate. The energy sector is no exception. Innovation in how we run our economies has been gradual over the past century, mostly incremental till the end of the twentieth century. With the development of renewable energies (REs) and their accelerated penetration of the grid, Utilities and Distribution Companies (DISCOMs) are brought back to the whiteboard charged with the challenge to develop a more agile and dynamic strategy that maintains their sustainability in the volatile, uncertain, complex, and ambiguous world around.

The world has now well understood that if we want to limit global temperature rise as per Paris Agreement, we have to look for alternate, cleaner sources of energy production. This has led many governments the world over to make REs a "Must Run" solution while gradually curtailing generation from conventional plants.

This chapter provides an overview of the developments in the Indian electricity sector since the year 1990 and describes the various challenges associated with the integration of increasing shares of renewable sources of electricity. The following section introduces different applications that advanced energy storage technologies can serve to make the Indian Electricity Grid more reliable. It also showcases examples from around the globe that can guide Indian DISCOMs on how the deployment of batteries can be an effective deviation settlement mechanism tool, can reduce penalties, provide for fast ramping, defer distribution transformer (DT) upgrades, and capture curtailments. In this chapter, we share insights into future developments and discuss potential business models which can make the integration of grid-based storage both technically feasible and commercially viable.

DOI: 10.4324/9781003281818-9

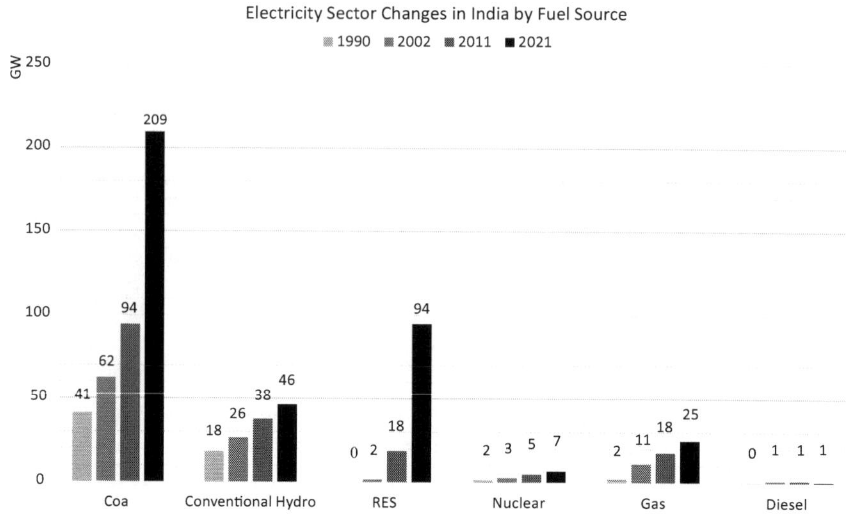

Figure 9.1 Changes in the resource composition of the Indian electricity mix, 1990–2021.

Source: CEA (2011, 2021)

9.2 Power Sector – Status Quo, Developments, and Projections

In March 2021, the generating capacity of the Indian power sector was 382.2 GW, roughly six times the capacity of the year 1990. All electricity sources contribute to this general growth trend, while the overall composition of installed electricity capacity is relatively dynamic over the years. In 2002, for example, coal-based generation stood at 59 per cent (62.1 GW) and RE sources at merely 2 per cent (1.6 GW), as depicted in Figure 9.1. In 2021, the share of coal in installed capacity has come down to 55 per cent (209.3 GW) and the share of new renewables increased to 25 per cent (94.4 GW) (CEA, 2021). As of November 2020, another 50 GW of renewable capacity is under execution and bidding.

India had a target of 40 per cent electricity generation capacity from non-fossil fuel sources – primarily REs – by 2030. Bloomberg New Energy Finance projects in its New Energy Outlook 2020 that 75 per cent of India will be powered by renewables in 2050 (BNEF, 2020a). This will pose major technical, institutional, and economic challenges for the sector, as described in more detail in the following section.

9.3 Electricity Sector Transformation Challenges

As shown in Figure 9.1, the RE penetration of the Indian grid has been increasing year on year. The Government of India has been ably supporting this growth by announcing renewables as a "Must Run" in early 2009, subjecting state DISCOMs to Renewable Purchase Obligations. Renewable sources of energy make an essential contribution to decarbonising the energy system. However, with higher shares of renewable energies in the power mix technical challenges are occurring, especially from grid management and grid expansion perspective.

9.3.1 Technical Challenges of RE Integration

Bound to the principle of reliability, utilities and DISCOMs consider the increasing penetration of renewables a risk. The intermittency of renewable sources of energy and their growing share in the electricity mix make the grid effectively more vulnerable. Projections show that by 2024 the Indian grid will carry around 75–100 GW of solar-generated electricity, which will result in steeper and longer ramps during sunrise and sunset hours. In other words, in the early mornings, the grid will have to handle a high level of solar generation ingress while electricity production from conventional base-load sources will get suppressed. Higher solar generation during the day would create challenges for these plants due to a load profile similar to the infamous California Duck Curve. During evening hours, when demand starts rising at around 5 pm, the fall in solar generation will have to be balanced off otherwise. The incongruency between renewable electricity supply and demand confronts the utilities with steeper and longer ramping blocks. As of date, the ramping rate is 150 MW/minute for 40 minutes. India Energy Storage Alliance (IESA) in-house estimates show that the ramping rate could go up to around 350MW/minute for 3–4 hours by 2030, which means there is an enhanced requirement of a balancing resource to be present in order to cater to this increased ramp requirement. Many of the existing conventional power plants are already running at below par utilisation factors and some even had to be shut down. As per government projections, 25 GW of thermal capacity will retire between 2020 and 2030. The displacement effect on conventional base-load fuels will gradually lead to a considerable reduction in spinning reserves – generation capacity that is online but unloaded and could compensate for generation or transmission outages within 10 minutes. Frequency-responsive spinning reserves would act even within 10 seconds to maintain system frequency. Spinning reserves are the type of reserve first used when shortfalls occur.

9.3.2 Institutional Constraints

The 2012 Indian Grid Blackout has given rise to many strict regulations to improve grid reliability. This includes stricter deviation settlement mechanism protocols for both generators and DISCOMs with high penalties for violating schedules, narrower frequency bands, and more stringent rules for upgrades and maintenance cycles. DISCOMs rely on software tools to determine RE generation. The variations between real-time renewable power generation and grid availability bring about inaccuracies that utilities find difficult to manage because of regularised forecasting and scheduling. Inadequate compliance by underestimating or overdrawing RE contributions is subject to penalties, ultimately resulting in revenue loss.

9.3.3 Other Issues

Further RE expansion and grid integration is perceived as rough and risky. The Indian Energy Industry has been facing several different hurdles, starting with anti-dumping duties whereby the import of Chinese modules is being discouraged by subjecting them to higher customs duties. At the same time, the local module manufacturing capacity is not developed at scale which forces investors and developers to rely on imports. A second issue is related to land acquisition, where oftentimes project-level challenges arise due to land falling under agricultural or forest land definition. A third issue is the poor financial health of DISCOMs, which leads to payment delays for the purchase of electricity from RE developers. A fourth problem with increasing grid penetration of renewables is the evacuation network which has not been expanded in unison and raises concerns in power curtailment, especially during months of high renewable resource influx. Fifth, the increase in rooftop solar photovoltaic (PV) in urban India has created problems of back-feeding at the DT level. One issue with DTs is their stationing in cities which makes upgrades an exhaustive and expensive task. Sixth, one should also not forget captive power generation, which, according to Central Electricity Authority (2021), stood at 55 GW at the end of FY2020–2021 and also relies on the grid while in turn increasing uncertainty to grid operations. There have been talks by Central Electricity Regulatory Commission and CEA about reducing load management blocks from presently 15 minutes to 5 minutes, which would result in increased frequency of load management dynamics going forward. Last, with the overall increase in electricity demand, partially as a result of the rise of the urban population, demand peaks being experienced across India have increased considerably. As per CEA Load Generation Balancing Report 2020–2021, India's electricity generation is anticipated to be 9.1 per cent peak surplus and 2.7 per cent energy surplus. From a regional perspective, only parts of South and East India experience regular but minor power deficits. All this together makes the journey towards 500 GW of renewables by 2030 a challenging task.

9.4 Technical Solutions

There are many technical solutions to the aforementioned concerns but only very few are technically, commercially, and environmentally viable. Thermal power plants are not designed to handle rapid ramping and load switching without significant deterioration in heat rate and increase in carbon dioxide emissions. Gas-based power plants cater better to the idiosyncrasies of REs, but cheap domestic gas is not available and many gas plants are being shut down or run as peaking plants. One technological solution that appears to fit better is pumped hydro, but concerns over capital costs and environmental viability make its prospects look bleak (CEA, 2020). Last, Distributed Energy Storage Technologies, including electrochemical batteries and thermal storage, are feasible alternatives. With around 85 per cent decrease in costs and continual improvements in technology, batteries are becoming a commercially more and more viable solution (IESA, 2020). The levelised cost of energy from batteries is now making a competitive bid against that of conventional power plants.

Both regulatory agencies and the industry have realised the importance of energy storage solutions for grid stability and reliability. As of 2020, India's total energy storage is 21 GWh. For the year 2027, the IESA forecasts a storage market of 37 GWh (base case) (IESA, 2020). If we look into application areas, around 68 per cent of the demand comes from behind-the-meter applications such as solar rooftop PV installations, electric vehicles, systems for uninterrupted power supply, and telecom towers, followed by grid applications. As a result of an increasing number of battery energy storage system (BESS) installations across the globe, the price has decreased by almost 87 per cent at the cell level (lithium ion) over the past 10 years. The price drop mainly attributes to the economies of scale achieved since the year 2015 for each manufacturer crossing the 5 GWh mark. At present, battery cell prices are available at around US$120/kWh and a further 30 per cent reduction is expected within the next two years. As per Bloomberg New Energy Finance, the battery cell price could be down to US$100/kWh by 2023 (BNEF, 2020b). Figure 9.2 shows how the price has fallen for battery packs from 2011 to 2019. According to Benchmark Mineral Intelligence (2024), the Li-ion battery cell price was down to US$ 96/kWh in December 2023. Joining the

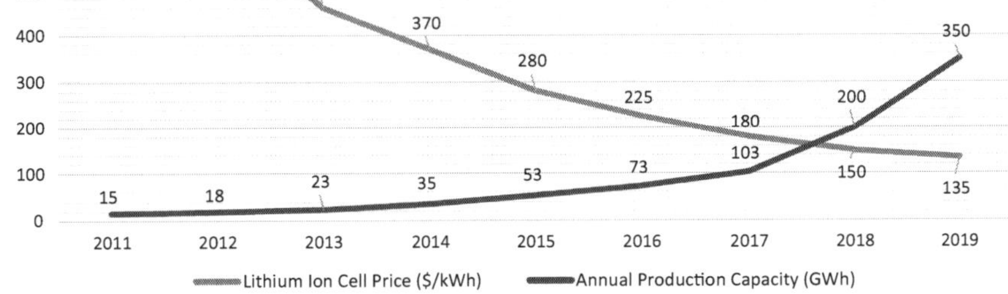

Figure 9.2 Battery price and manufacturing capacity trend (CES analysis).
Source: IESA (2020)

global trend, the Government of India announced an INR 18,100 crore (US$2.5 billion) scheme for advanced chemistry cell battery manufacturing in India linking the incentive to output-based local value creation.

9.5 The Unique Opportunity of Wind and Solar Resources

Wind and solar resources are complimenting each other. Generally speaking, during the daytime, when the solar irradiance is high, wind speed is relatively low. By contrast, during the early morning or late evening hours when due to the non-availability of sun, solar generation plummets, the winds are speeding up due to the arising temperature difference. During monsoons, wind speed generally remains high throughout the day but reduces towards evening hours, whereas solar is highly fluctuating and low during this time. In fact, during February and March when there are almost no cloud covers, solar is much less variant. This complementing nature of wind and solar resources can be utilised, and this alone solves many issues in the renewables industry. If we combine wind and solar in one plant, they together can optimise land utilisation, the overall economics due to shared infrastructure, and the issue of intermittencies.

Against this backdrop, hybrids have been gaining importance over time. As of date over 50 hybrid projects are announced worldwide with Australia and United States leading in installations. Simply put, a hybrid system can combine wind and solar with an additional resource of generation or storage. This in many ways solves the issue of RE volatility and the goal of round-the-clock power supply from renewables. When storage is added, this also makes renewables more reliable. The storage of energy can be through various forms, electrochemical, pumped hydro, gas Peaker plants, gravity storage, flywheels, etc. Pumped storage is preferred for longer durations of storage but is highly underexplored due to exploration costs and infrastructural challenges. India has a potential of 96 GWs of pumped hydro, with presently only 4.8 GW operational and 1 GW under construction. Electrochemical solutions, especially lithium-ion batteries, by contrast, are more economical for shorter durations of four to six hours.

9.6 The Economics of Storage Integration

The importance of energy storage for large-scale RE grid integration is well understood and largely acknowledged in today's energy market. However, back in 2017 when a BESS model was deliberated upon for India, the economics pointed at a high CAPEX requirement for the complete integration of a BESS unit, including additional generation incurred by the storage component. One storage unit effectively handling grid curtailment over a duration of four hours would add INR 10/kWh to the electricity tariff (as compared to an RE-only tariff of INR 2/kWh). The major barrier to storage deployment is that multiple issues need to be addressed to make BESS economical. That is mainly due to the seasonality of renewables. In the case of wind, for example, the high wind season in India ranges mainly between June and October, and during these months wind assets do not need to be curtailed every day. Battery storage applications thus find a rather small deployment window and would remain unutilised throughout the rest of the year. Therefore, it is much required that storage deployment finds multiple applications. For wind, this could be (1) forecasting and scheduling management by which to reduce penalties imposed by the deviation settlement mechanism, (2) minimising curtailment, and (3) catering to ancillary markets. These different battery applications can be utilised throughout the year and guarantee continuous revenue streams or savings. In the case of solar, battery storage finds more applications and is easier to integrate because the variability is less and more storage is needed due to regular non-solar hours. In other words, at the start, a BESS makes more economic sense in combination with solar power generation but later also in wind-based power generation.

Over the past decade, the price of lithium-ion batteries has dropped by 87 per cent. What started off at above US$1,000/kWh has come down today to roughly US$150/kWh. This trend is expected to reach a price of US$100/kWh and below by 2027. However, the cost of a storage system hinges upon a number of factors such as round-trip efficiency, the number of cycles that the battery runs in its lifetime, the depth to which the battery is discharged, and also the surrounding temperature and other conditions. If a battery is discharged fast or frequently, this has an effect on its life and also makes the battery more expensive compared to a slow charge-discharge cycle. In principle, the price of a stand-alone battery system is much higher than a collocated project of renewables and battery storage. This is primarily because of the cost of common infrastructure-related the balance of storage (BoS) system getting shared. Also, with an increase in project size and duration, the overall cost comes down. The BoS cost, for example, reduces significantly for a four-hour collocated storage system compared to a one-hour collocated system.

The overall project level breakup of a storage system cost has decreased a lot over the past few years. Along with that, storage integration and the hours of storage are considered to have increased. In the United States, the tariffs for renewables with storage saw a decrease of 15 per cent in between power purchase agreement prices of 2018 and 2019 (LBNL, 2020). According to a Customised Energy Solution (CES) internal analysis, the predominant share (65 per cent) of a typical cost breakup of a storage project with four hours deployment constitutes the BESS with Battery Monitoring System followed by Power Control System and Energy Management System (16 per cent), the container (10 per cent), the BoS system (7 per cent) and Project Management, and five-year operation and maintenance at 1 per cent each.

As discussed earlier, a combined REs and battery storage system is more beneficial for the duration of storage up to four hours. It is expected that the percentage of storage will be around 10, which slowly will increase to around 50 per cent as the price of batteries comes down. In a best-case scenario with

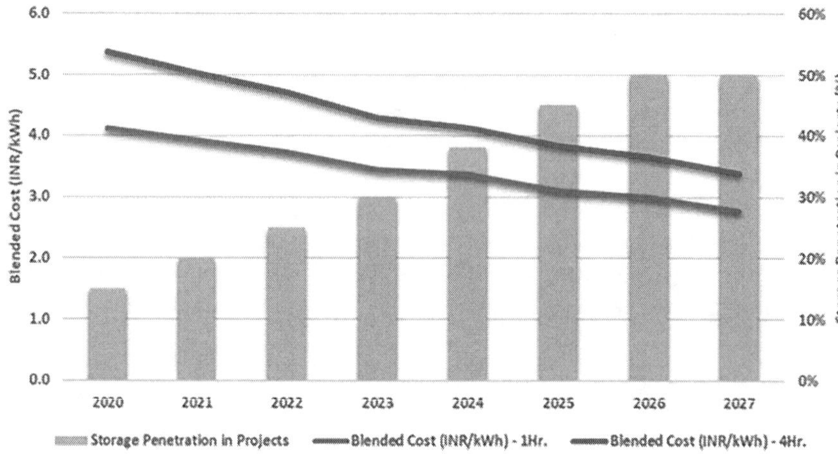

Figure 9.3 Blended cost trend (RE + storage).
Source: IESA (2020)

proactive policy support, that is tender authorities releasing more hybrid tenders with storage components, and the vision of 450 GW being realised by 2030, storage capacity in 2027 could be around 12 GWh. In a more conservative scenario, storage capacity could be 5 GWh by 2027. The tariff adder based on today's cost for four hours storage is INR 7/kWh. This is expected to reduce to INR 3/kWh by 2027, as can be seen in Figure 9.3. The projected tariff addition for one hour storage will be around INR 6/kWh, down by 50 per cent from today's INR 12/kWh.

The blended cost of renewables plus storage, however, is not a linear tariff addition. The blended cost also varies depending upon the size of storage. The size of storage, in turn, hinges upon the amount of charging required from the RE system. In other words, tariffs are changing with more renewable energies to be stored in batteries. Based on today's prices and considering technological advancements for both renewables and storage, this cost is expected to drop and the tariff accordingly. At present, in a 50 per cent penetration scenario, the blended cost for a four-hour system can range between INR 5 and INR 8/kWh. This is expected to drop to around INR 2–4/kWh by 2027.

9.7 Storage as Enabler for Utilities and DISCOMs – Case Studies

9.7.1 Energy Storage for Transmission Investment Deferral

The RE addition into the Indian power grid is expected to cross 175GW by 2022, of which 160 GW comes from solar and wind energy resources and 120 GW is of utility-scale projects that are directly connected to the transmission grid network (NITI Aayog, 2019). These two renewable sources of power are – as is known – variable in nature. The typical total peak energy from solar and wind combined crosses 100 GW only for 29 days a year for 4 hours per day. This constitutes merely 1.3 per cent of the total time of a year. Most of these decentralised wind or solar farms require transmission capacity additions or transmission line expansions. As per the National Electricity Plan Vol. II, 47 GVA of transmission addition is necessary with the national RE target hitting 175 GW. An internal analysis by CES concludes that when RE generation crosses 100 GW only 1.3 per cent of the time, transmission investment can be saved with the installation of up to 5 GW/15 GWh battery energy storage. Moreover, many wind and solar farms are more than 50 km away from the existing transmission structure. Nearly 50 per cent of the projects are well within this range of 50 km. However, by installing an energy storage system (ESS) to store excess generation, there can be significant savings on transmission line investments. According to an IESA internal analysis (2020), the Indian energy storage market has a potential of 300–700 MWh in 2022 going up to 3,000–4,900MWh in the year 2026. Hence by avoiding power from being transmitted across lines, there can be savings on the transmission cost infrastructure of about 6 GVA of substations and transmission line capacity. System modelling studies can help determine the optimal location for energy storage, which can be also used for alternate applications such as RE firming and RE output smoothening. All this is conducive to multiplying the overall benefits while delivering reliable power from renewable sources to the consumer at the lowest possible price.

In India, RE curtailment is a major issue due to power evacuation challenges. Besides, key challenges faced by developers are the CAPEX involved in transmission capacity enhancement, obtaining the right of way, and land clearances for substations. ESS is an immediate remedy for such cases, with a much lower gestation time. Data generated during the high wind capacity months at a wind farm in Rajasthan show that due to evacuation constraints the independent power producers (IPPs) were losing on generation. The windfarm was operated in the curtailed mode, which resulted in lower generation and thus revenue loss. Such losses can be avoided by storage installation while making much more business sense to address curtailment loss than a total transmission network revamp.

9.7.2 *Energy Storage From a Distribution Perspective*

Distribution grids in India have a lot of issues. There are power quality issues due to low power factor (PF) and harmonics, low network reliability, and high technical and commercial losses. The 415 V and 11 kV networks are considered not flexible enough for voltage and power quality fluctuations, as automated on-load tap changers are used mostly at 33kV networks and above. In such a scenario, higher deployment of rooftop solar PV, as detailed in Figure 9.4; integration of EV chargers; and rising commercial loads can

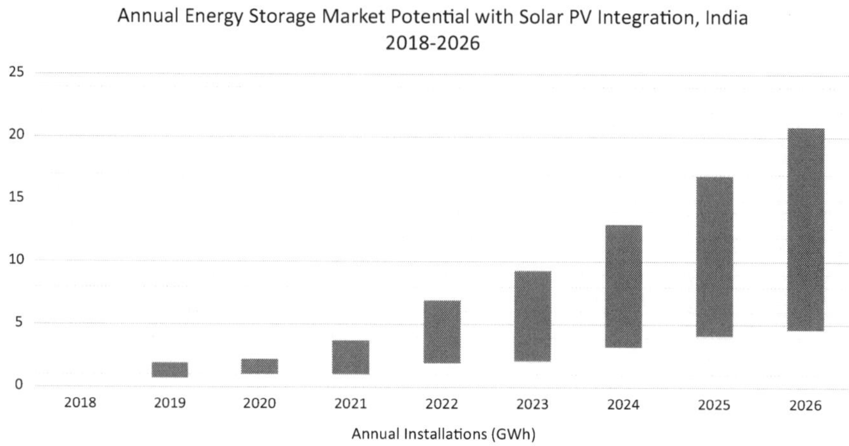

Figure 9.4 Annual energy storage market potential (solar PV integration).
Source: NITI Aayog (2019)

create a lot of difficulties for the operators in maintaining network stability. As witnessed across the world, ESS can provide the missing flexibility for low- and medium-voltage electricity networks.

BESSs have got three advantages to offer at feeder/DT level that wants to facilitate solar PV integration. They can defer distribution upgrades, have peak shaving benefits, and can maintain power quality through voltage regulation, as discussed in more detail in the following sections.

9.7.2.1 Distribution Upgrade Deferral

ESSs can help to defer distribution system upgrades by a few years depending on load and distributed generation growth rates. The benefits of an ESS are calculated considering savings on interest payments for upgradation costs of cables and substation equipment that result from the deferral. The deferral case works really well when afternoon load on the network is low and/or load growth on the network is low on average within a year. Hence, saturated areas like metros are the ideal application case for storage. Also, cities, where substation upgrades are not possible due to space constraints, are suitable for such deployment.

9.7.2.2 System Peak Shaving Benefits

ESSs can capture peak shaving benefits. ESS can be deployed to reduce utility loads during system peaks – mostly during evening hours. In reverse, during low-demand hours, storage can be used to capture excess solar production which can get used during non-solar hours.

9.7.2.3 Power Quality Savings – Voltage Regulation Perspective

Feeder-level load is having a regular PF of 0.9. When solar is penetrating the grid, the PF reduces to around 0.8. This reduction is the result of solar feeding in primarily active power which changes the ratio of apparent power to active power. Storage solutions can help to maintain the baseline PF value thus ensuring smooth operations.

CES conducted a study for the Indian Smart Grid Forum, dedicated to the development of an ESS distribution grid road map for Calcutta Electric Supply Corporation (NITI Aayog, 2019). As part of the study, a few load centres in Kolkata were analysed, which showed annual peak load growth of 3 per cent over the past five years. The introduction of lithium ion BESS at varying capacities was studied for the years 2020, 2022, and 2025. ESS capacities in the range of 10 per cent to 20 per cent of the DT capacity in these load centres were found to defer the distribution investment by up to four years plus respective internal rate of return (IRR) on ESS services. Deferral, apart from energy shifting and PF improvement, was one of the key effects. However, on a case-by-case basis, different applications were found to drive the need for ESS at the distribution utility level.

The CES study yielded the following findings:

- To get a maximum IRR, 20 per cent of storage for DT capacity in the years 2022 and 2025 is ideal. This storage option is also suitable for 2020 at the price of negative IRR but with still high deferral benefit.
- Maximum IRR is driven by low storage cost.
- For higher solar penetration (up to 70 per cent in 2025), storage still provides economically viable results, given its appropriate size and technology.
- Leading metro cities in India showed peak load growth between 2 and 7 per cent in 2018. For the years 2019–2026, various factors such as the adoption of energy efficiency measures and practices by residential, commercial, and industrial customers as much as space saturation may lead to lower peak demand growth in these cities.
- Peak load growth of key cities such as Delhi, Mumbai, Kolkata, and Bengaluru is expected at 3–4 per cent annually during 2019–2026. Such cities with a high concentration of commercial activities, estimated to be 10–20 per cent of the total city load, could witness growth 1–2 per cent lower per year. Peak load centres may grow at a slow pace of 2–3 per cent annually.
- In this case, DISCOMs will be discouraged from investing in distribution grid upgrades or expansions due to low returns and longer return on investment.

ESSs make a good case under such conditions and can score with their flexibility. If the peak load remains flat after a few years of operation, the ESS unit can be disassembled and simply shifted to another location.

9.7.3 Storage for DISCOMs

To make the grid more reliable and attain energy security, a deviation settlement mechanism was introduced in 2014 and since then amended five times till date (IESA, 2020). Penalties for not adhering to the schedule have become strict. The average price of electricity transacted through the deviation settlement mechanism is around INR 3 per unit but can go up to a maximum of INR 8 per unit. With the new vector, the deviation settlement mechanism penalty can go up to even INR 16 per unit in case of violation of the band. The charges of the deviation settlement mechanism are unavoidable, but they can be optimised through smart grid solutions. In fact, zero deviation is a sheer coincidence rather than the expected case. Battery energy storage solutions are a prime alternative for short-term balancing in order to reduce the price of prediction uncertainties.

In its present form, the operational frequency band has been tightened to 49.85–50.05 Hz. The step size has been reduced to 0.01 Hz. The maximum allowable deviation is restricted to 12 per cent of the schedule and penalties are narrowly defined. With more renewables penetrating the grid, it will become more difficult for DISCOMs and utilities to adhere to schedules. This is where energy storage can make a major contribution, especially during time blocks when utilities draw energy above the overdraw limit. When the penalties incurred are more than the cost of power from batteries, the potentially overdrawn amount of energy can be met by discharging the batteries. Similarly, in reverse, when utilities draw less than the allowable limit, the excess can be used to charge the batteries. In this way, utilities can save on paying penalties due to over- or under-drawings of the energy band. One more application where storage can work well for distribution utilities is in demand response services. This market is not developed in India yet, but significant growth can be expected.

9.8 Conclusion

Energy storage is no longer a mere option but has become a necessity for the energy sector to sustainably develop further. On-the-ground projects have further proved the fact that storage projects are viable not only technically but also commercially. The way forward will see further mainstreaming of the technology for applications across the value chain in a more efficient and economic manner.

Overall, ESS appears to be the best solution for the massive expansion of renewables in the Indian grid. With land availability and acquisition being a major challenge, utilising the available land to the fullest through hybrid solutions makes a lot of sense. With the government releasing the National Renewable Energy Hybrid Policy (MNRE, 2018) and recent developments like removing tariff ceilings, reduction in the cost of storage, advancements in inefficiencies, the forecast for hybrid projects in India looks good. Such projects work much better with a two-part tariff structure. These projects help attain better project economics and increase the participation of developers and IPPs in upcoming bids. The tariff trend also helps RE-cum-storage projects as more and more renewables penetrate the grid. Thermal assets are gradually being run on part loads which is ultimately increasing their per-unit generation cost due to higher heat rate and lower efficiency which results in plant load factors of 55 per cent on average across units. Talks have also started over lowering this minimum set point of coal units, which will further increase the per-unit cost of generation. Moreover, the cost of installing Flue gas desulphuriser and more improved emission control and monitoring equipment is making new coal plants and similarly gas peaker plants more expensive than RE hybrids. By contrast, for RE hybrids, the cost trend will only decrease even with technological improvements. Therefore, the future of hybrids is bright and will make renewables mainstream. The world around us is ever-changing and innovative. Change has become normal and to improve on a day-to-day basis is to maintain one's competitive edge in the market. Hybrid systems provide the much-required inflection point which the renewables industry needs to sustain its growth phase. Finally, the future of technology looks promising with many options congregating in the power sector. It is a challenging and competing environment in which the technology that remains agile, compatible, convenient, and customer-oriented will sustainably assert its lead.

References

Benchmark Mineral Intelligence (2024). *Lithium-Ion Battery Cell Price*. Retrieved from: https://www.benchmarkminerals.com/lithium-ion-battery-prices/?tab=chart (Last accessed: 21.01.2024)

BNEF (2020a). *Bloomberg NEF's New Energy Outlook 2020*. Bloomberg New Energy Finance. October 2020. Retrieved from: https://about.bnef.com/new-energy-outlook-2020/ (Last accessed: 19.08.2021)

BNEF (2020b). *Bloomberg NEF's Annual Battery Price Survey*. December 2020. Retrieved from: https://about.bnef.com/blog/battery-pack-prices-cited-below-100-kwh-for-the-first-time-in-2020-while-market-average-sits-at-137-kwh/ (Last accessed: 19.08.2021)

CEA (2011). *Annual Report 2010–11*. Central Electricity Authority. June 2011. Retrieved from: https://cea.nic.in/wp-content/uploads/2020/03/annual_report-2011.pdf (Last accessed: 21.05.2021)

CEA (2020). *Report on Optimal Generation Capacity Mix for 2029–30*. Central Electricity Authority. January 2020. Retrieved from: https://cea.nic.in/old/reports/others/planning/irp/Optimal_mix_report_2029-30_FINAL.pdf (Last accessed: 13.08.2021)

CEA (2021). *Executive Summary on Power Sector*. Central Electricity Authority. March 2021. Retrieved from: https://cea.nic.in/wp-content/uploads/executive/2021/03/executive.pdf (Last accessed: 21.05.2021)

IESA (2020). *India Stationary Energy Storage Market Overview Report – 2020–2027*, 6th edition. India Energy Storage Alliance. July 2020. Retrieved from: https://indiaesa.info/resources/industry-reports/3106-india-stationary-energy-storage-market-overview-report-2020–2027 (Last accessed: 25.05.2021)

LBNL (2020). *Estimating the Cost of Grid-Scale Lithium-Ion Battery Storage in India*. Lawrence Berkeley National Laboratory. April 2020. Retrieved from:

https://eta-publications.lbl.gov/sites/default/files/lbnl-2001314.pdf (Last accessed: 30.08.2021)

MNRE (2018). *National Wind and Solar Hybrid Policy*. Ministry of New and Renewable Energies. May 2018. Retrieved from: https://mnre.gov.in/img/documents/uploads/2775b59919174bb7aeb00bb1d5cd269c.pdf (Last accessed: 19.08.2021)

NITI Aayog (2019). *Energy Storage System – Roadmap for India 2019–2032*. ISGF, IESA in Collaboration with McArthur Foundation. August 2019. Retrieved from: http://niti.gov.in/sites/default/files/2019-11/ISGF.pdf (Last accessed: 01.06.2021)

10 A Realistic Assessment of the Prospects of Biofuels for Transportation in India

Abhijit Banerjee

10.1 Introduction

India is the third-largest consumer of primary energy in the world after the United States and China (MoPNG, 2020). Being one of the fastest-growing economies in the world, its share of global primary energy is expected to nearly double from around 6 per cent in 2019 to 11 per cent by 2040 (BP, 2019). Like in other emerging economies, the share of transportation in total final energy use has been rising steadily, from 16 per cent in 2005 to 21 per cent in 2020 (Gupta et al., 2020). Among fuel sources, petrol and diesel together make up more than 90 per cent of supplied energy, while road transport accounts for 90 per cent of the energy used in the transport sector (IEA, 2020). If current trends continue, the energy demand for the road transport sector may double by 2040 (IEA, 2021). Of great concern is the fact that India is highly dependent on petroleum imports, gaining the status of the third-largest oil importer in the world. Imports make up around 80 per cent of its petroleum demand, a proportion that may increase in the future if demand rises faster than domestic production (Aradhey, 2019). Petroleum has by far dominated India's import bill, representing 17 per cent of the import share in 2019 (Chandra, 2020).

In line with growing global interest in the purported benefits of biofuels, India started adopting policies to promote biofuels in the early 2000s. The primary rationale for India's biofuels programme was the reduction of oil imports widely promoting an aspirational goal of 10 per cent import reduction by 2022 (Das, 2020). Yet, India's crude oil imports reached a record high in December 2022 as per OECD data (CEIC, 2023).

Desired secondary benefits included the alleviation of environmental burdens and the promotion of rural development. Environmental benefits most commonly associated with biofuels, under the right circumstances, include the reduction of vehicular air pollution and greenhouse gas (GHG) emissions. In the first decade of the twenty-first century, reduction of air pollution was arguably the bigger concern in India; this continues to be the case with many Indian cities ranked among the worst globally in terms of air pollution (Broom, 2020). India's Nationally Determined Contributions (NDCs) for the

DOI: 10.4324/9781003281818-10

2015 Paris Climate Agreement included a pledge to reduce the carbon intensity of gross domestic product (GDP) by 33–35 per cent by 2030 (MoEFCC, 2021). The revised set of NDCs contains a sharpened goal of even a 45 per cent reduction in the carbon intensity of GDP by 2030. The increased use of biofuels for transportation may contribute to that goal.

This chapter explores India's experience with liquid biofuels and the evolution of its biofuels policies from the early 2000s till date. Despite early setbacks, the Government of India (GoI) has continued to push for biofuels development and adoption, with policy modifications and new directions. The main objective of this chapter is twofold: first, to determine whether liquid biofuels can contribute significantly to India's transportation sector and, second, to examine potential negative consequences of biofuels development and if these are receiving due consideration in policy documents. The research relies on secondary literature but tries to "connect the dots" and draw "big-picture" inferences, without delving into details of technology or policy minutiae. While ample literature exists on this topic – from academic scholarship to reports by governments, think tanks, and international organisations – most are focused on limited aspects such as technical potential, economic viability, or policy critique of biofuels in India. This chapter, by contrast, takes a "bird's eye view" to analyse the potential and constraints, while considering direct and indirect economic, social, and ecological consequences at the local and global scales.

The chapter is organised into six parts. The following section 10.2 puts biofuels development in the global context, briefly highlighting key trends and challenges. Section 10.3 traces the evolution of biofuels policy in India, noting important developments and influences. Sections 10.4 and 10.5 delve deeper into ethanol and biodiesel development trends and prospects. Finally, Section 10.6 concludes with an argument for a more holistic approach to sustainable biofuels incorporating the entire transportation sector and a more thoughtful consideration of indirect consequences.

10.2 Biofuels in the Global Context

Since petrol and diesel are the most important liquid transportation fuels, global efforts to find biofuels that can be blended with petrol (ethanol) and diesel (biodiesel) started in earnest in the 1990s. Brazil has the oldest commercial-scale bio-ethanol programme in the world going back to the 1980s, while the United States started large-scale ethanol production in the 1990s. Today, the United States and Brazil are by far the biggest bio-ethanol producers in the world accounting for 55 per cent and 30 per cent of global production, respectively (EIA, 2019). In terms of ethanol feedstocks, Brazil uses sugarcane, while corn (maize) is the most important source in the United States. Biodiesel production at a commercial scale started only in the first decade of the twenty-first century. Today, the most significant biodiesel players globally include the European Union (EU) (35 per cent share of global production), followed by Indonesia (17 per cent), and the United States (14 per cent) (EIA, 2019). For biodiesel, the most important feedstocks used include oils derived from soybean, palm, rape, and sunflower.

Many countries have adopted mandates for biofuels promotion, typically coupled with supportive policies. In the United States, the renewable fuel standard (RFS), administered by the Environmental Protection Agency (EPA), was launched in 2005 and sets use targets for biofuels that are ratcheted up over time (EPA, 2021). For bio-ethanol, about 10 per cent blending with gasoline (E10) has been achieved in recent years, although there are doubts whether this share can be increased (Bracmort, 2020). The share of biodiesel has not even reached 5 per cent overall, although different blends are available in different parts of the country (EIA, 2020). In the EU, the Renewable Energy Directive (RED I) of 2009 set a goal of 10 per cent of transportation energy to be derived from renewable sources by 2020 (EC, 2020). A share of just above 8 per cent renewable energy in transportation (not only biofuels) was achieved in 2018, while the fuel blending volume is about 5 per cent overall, though individual countries may have higher shares (Panoutsou et al., 2021). Brazil's sugarcane-ethanol programme was instituted in response to the oil shocks of the 1970s and has been spectacularly successful over the decades in terms of reducing oil imports. Blending mandates have been gradually strengthened over time, which currently is 27 per cent for gasoline. However, the actual blending rate is somewhat higher, since a small but significant number of vehicles can run on even 100 per cent ethanol (E-100) (TP, n.d.).

Since biofuels are derived from biomass, it is expected that GHG emissions reduction will be accomplished when they fully substitute petroleum-based fuels. However, large-scale cultivation of "energy crops" – whether corn, soybean, sugarcane, and the like – has been linked with many direct and indirect negative environmental consequences. Most obviously, large-scale industrial agriculture is associated with intensive water use, chemical use leading to water pollution, as well as soil erosion (NRC, 2011; EPA, 2018; Jeswani et al., 2020). Aside from these local impacts, indirect land-use change (ILUC) and pressure on food prices have become major global concerns associated with energy crops. When energy crops displace food crops in one part of the world, food production moves elsewhere. In addition, lucrative export markets for biofuel feedstocks lead to the rapid expansion of energy crop plantations in poorer countries. The combined impact of these two factors has been documented to have caused large-scale deforestation in many critical biodiversity hotspots such as Brazil, Indonesia, and Malaysia. When GHG emissions from ILUC are included in the calculation, most biofuels derived from such energy crops show no net GHG benefit or come out even worse than petroleum-based fuels (De Oliveira et al., 2005; Goldemberg et al., 2008; Mosnier et al., 2012; EPA, 2018; Jeswani et al., 2020). Diversion of land from food crops to energy crops also creates upward pressure on food prices since food commodities are traded globally. In a food-insecure world, the food-versus-fuel dilemma has been labelled "morally repugnant" by human rights advocates (Tenenbaum, 2008; Oxfam, 2008; FAO, 2013; Renzaho et al., 2017; Subramaniam et al., 2019).

As the undesirable consequences of energy crops have become better well known, there have been increasing calls for limiting such "first generation" (1G) biofuels derived from energy crops. The so-called second generation (2G) biofuels can, in theory, be produced from waste or residue and thus avoid the problem of dedicated land for energy crops entirely. "Cellulosic" 2G ethanol can be derived from agricultural or forestry residues, while "advanced" (2G/3G) biodiesel can be derived from municipal waste, used cooking oil (UCO), animal fat, or algae. However, 2G/3G biofuel production is technologically complex, and combined with supply chain constraints, their commercial viability at scale is not yet established, despite decades of research and government support (Flynn, 2017; Padella et al., 2019; IRENA, 2019; Panoutsou et al., 2021). In the United States, the RFS originally called for more than 10 billion gallons of cellulosic ethanol by 2020; the EPA had adopted a reduced 2020 target of 590 million gallons (Bracmort, 2020), but actual production was less than 20 million gallons (Nuelle, 2019). In the EU, biodiesel dominates the biofuel sector (80 per cent share) and the EU is a huge importer of palm oil feedstock from Indonesia and Malaysia, where the expansion of palm oil plantations has fuelled large-scale deforestation over the past years (Valin et al., 2015; Vijay et al., 2016; IUCN, 2018; Lustgarten, 2018; Greenpeace, 2018; WWF, 2021). In response to growing criticism, the EU tried to adopt sustainability criteria for biofuels, but experience shows that certification schemes have not made a meaningful difference (Stattman et al., 2018). Sustained campaigns led the EU to adopt an updated RED II in 2018, which freezes crop-based biofuels at 2020 levels while phasing out support for high-deforestation-risk biofuels by 2030. RED II also sets targets for "advanced" biofuels, but critics give to consider that other problematic feedstocks such as soy oil should also be phased out (T&E, 2021).

10.3 The Evolution of India's Biofuels Policy

India's adoption of biofuels has been motivated by the desire to reduce oil imports and environmental pollution while benefitting the agriculture sector with a vision to gradually increase the blending of ethanol and biodiesel in petrol and diesel fuels. After completing successful field trials in 2001, the GoI launched the Ethanol-Blended Petrol (EBP) Programme in 2003 for the sale of 5 per cent EBP (E5) in selected states. This was expanded to most of the country, except for some Himalayan states, from 2006 onwards. Oil marketing companies (OMCs), most of them in the public sector, were required to sell E5 "subject to commercial availability" (Purohit & Dhar, 2015; Sarwal et al., 2021). The National Biodiesel Mission (NBM) was launched in 2003 with the goal of achieving 20 per cent blending in diesel (B20) by 2012. However, in 2009, the newly formulated National Policy on Biofuels (NPB-2009) pushed non-mandatory blending targets for both ethanol and biodiesel at 20 per cent to 2017 (Purohit & Dhar, 2015; Das, 2020). The objective of the NPB-2009 was to facilitate the development of the entire biofuel value chain in a comprehensive manner, including feedstocks, production, and blending through the adoption of an appropriate policy framework. To avoid the "food vs. fuel" dilemma, non-edible crops and feedstock were considered to be the explicit focus for biofuels development (Beckman et al., 2018).

In the years immediately following the launch of the EBP in 2003, the actual availability of ethanol remained very limited, in terms of both quantity and geography, with overall national blending rates barely reaching 1 per cent before the launch of NPB-2009. Even after the NPB-2009 was introduced, it took until 2016 for the national blending rate to reach 3 per cent (Gupta et al., 2020; Sarwal et al., 2021). Clearly, the 20 per cent ethanol blending target by 2017 was quite unrealistic. The main feedstock for ethanol in India has been sugarcane molasses, the dark, viscous liquid left after extraction of sugarcane juice. Ethanol has established uses – industrial and potable – in India, and the low price offered by the OMCs administering the blending programme was seen as the most important factor limiting the supply of ethanol for the EBP. A burdensome interstate transport tax regime also limited the movement of ethanol from sugarcane/ethanol surplus states to other parts of the country (Purohit & Dhar, 2015; Aradhey, 2019). Acknowledging the limitations, more favourable ethanol purchase prices were set by the GoI in 2014–2015. Since then, ethanol supply and blending rates have improved gradually (Sarwal et al., 2021), as discussed in detail in Section 10.4.

The main focus of the NBM was the development of biodiesel from non-edible oil seeds such as Jatropha and Pongamia, in order to avoid competition with food crops. In the first decade of this century, there was worldwide hype about the potential for plants like jatropha to produce enormous amounts of biodiesel, and India was no exception to this trend (Kant & Wu, 2011). *Jatropha curcas* is a "weedy" plant common in tropical and sub-tropical regions whose seeds produce copious amounts of oil that can be refined into biodiesel suitable for diesel engines. The fact that jatropha plants can grow in poor quality soil without many agricultural inputs, including water, seemed like a huge advantage (Biswas et al., 2010; Kant & Wu, 2011). Indeed, jatropha promised "sustainable" biofuels without food conflicts and with an increase in rural incomes as a bonus! An ambitious target of 10-plus million hectares of jatropha plantations, mostly on unproductive or "waste" lands, was adopted in the 11th Five Year Plan (2007–2012) by the Planning Commission of India. Many state governments also launched their own schemes, often with incentives for agro-business. Even the Indian Railways jumped on the jatropha bandwagon with plans for substantial jatropha plantations on surplus railway land (Ministry of Railways, 2003; Biswas et al., 2010; Kant & Wu, 2011).

However, it quickly became apparent that jatropha plantations were not commercially viable in India. The most fundamental problem was that proponents of jatropha had significantly overestimated seed production rates under low-input, low-fertility conditions (Biswas et al., 2010; Kumar et al., 2012; Ganguli et al., 2018). Seed production under intensive, high-input conditions remains unknown since previous cultivation experience is virtually non-existent; more importantly, intensive, high-input cultivation would defeat the original aim of the biofuels programme to avoid conflict with land for food crops. In fact, there is considerable evidence that in the mad rush for jatropha adoption, plantations were not restricted to unproductive "wasteland" but often expanded on land used for food and livestock production (Ariza-Montobbio & Lele, 2010; Findlater & Kandlikar, 2011). After a long gestation period of several years for plant maturity, when commercially significant amounts of seeds were not produced, both small farmers and agro-businesses lost money, leading to bitter disappointment and resentment (Ariza-Montobbio & Lele, 2010; Ganguli et al., 2018). Commercially significant amounts of biodiesel were never produced from jatropha, at least not on a national scale, and biodiesel producers mostly switched to other feedstocks to remain in business. Other countries investing in jatropha had similar experiences and many observers have just labelled it a "fiasco" (Kant & Wu, 2011).

Taking into account lessons from past experience as well as recent developments, the GoI came out with a new NPB in 2018 (NPB-2018). The NPB-2018 set new blending targets of 20 per cent for ethanol and 5 per cent for biodiesel to be achieved by 2030. In addition, there is a 10 per cent intermediate blending target for ethanol by 2022 (Aradhey, 2019). To achieve these targets, a much wider variety of feedstocks are considered, with special emphasis on "advanced" (2G and 3G) biofuels. A more comprehensive policy framework and incentives are also envisaged (Das, 2020). These revised targets are arguably more realistic, but significant challenges remain. The challenges and prospects for ethanol and biodiesel in light of the NPB-2018 will be analysed separately in Sections 10.4 and 10.5.

10.4 Ethanol

As noted in Section 10.3, ethanol supply for blending with petrol has been increasing gradually with more attractive purchase prices offered by the OMCs since 2014–2015 (see Table 10.1). The NPB-2018 added other favourable policies such as long-term purchase agreements, financial benefits such as reduced Goods and Services Tax, low-interest loans, interest subvention, and accelerated depreciation, and easier interstate shipment of ethanol. It also expanded the type of feedstocks from which ethanol production will be permitted (MoPNG, 2020; Chandra, 2020; Sarwal et al., 2021). As a result, ethanol production for blending has kept increasing over the past few years, hitting 5 per cent in 2019 (see Table 10.1). On the back of a bumper sugarcane crop in 2020, ethanol production for blending jumped significantly, and it is expected that the blending ratio will cross

Table 10.1 Ethanol supply for blending and blending ratios in India.

Year	*Quantity supplied (billion litres)*	*Blending percentage by OMCs*
2013–2014	0.38	1.53
2014–2015	0.67	2.33
2015–2016	1.11	3.51
2016–2017	0.67	2.07
2017–2018	1.51	4.22
2018–2019	1.89	5
2019–2020	1.73	5
2020–2021 (est.)	3+	8+

Source: Sarwal et al. (2021)

Table 10.2 Projected petrol demand and ethanol blending targets in India.

Year	*Petrol demand (billion litres)*	*Ethanol quantity required in billion litres*		
		5% blending	*10% blending*	*20% blending*
2021	39	1.95	3.9	7.8
2025	49.4	2.47	4.94	9.88
2030	57.9	2.90	5.79	11.58

Source: Sarwal et al. (2021)

8 per cent in 2020–2021, which was also helped in part by lowered petrol consumption due to the lockdown induced by the COVID-19 pandemic (Chandra, 2020).

Although the big jump since 2020 is the result of an unusual bumper crop, the increasing supply trend has raised the tantalising prospect of meeting the 10 per cent blending target, if not in 2022, at least soon thereafter. However, the 20 per cent blending target for 2030 still appears challenging in light of the steady increase in petrol demand based on recent trends. The Working Group on Enhancing Refining Capacity constituted by the Ministry of Petroleum and Natural Gas (MoPNG) as cited in Sarwal et al. (2021) has assumed year-on-year demand growth for petrol at 3–4 per cent (see Table 10.2). Based on these projections, the required quantities of ethanol for different blending targets are presented in Table 10.2.

It is clear that a 5 per cent blending target can be met for the foreseeable future, unintended consequences aside, even when taking into consideration the increasing petrol demand. However, meeting the 10 per cent and 20 per cent blending mandates will require a near-doubling and a nearly fourfold increase in ethanol production just for blending, from a 2019–2020 base year. The prospects for increasing ethanol supply for blending from 1G versus 2G sources are discussed in the following subsections.

10.4.1 Prospects for 1G Ethanol

The main feedstock of ethanol production for blending purposes has so far been sugarcane molasses. However, it is important to recognise that sugarcane molasses have useful applications in other industries, especially in the food and beverage sector. Unless sugarcane cultivation increases substantially, there is a shortage of ethanol production from molasses (Purohit & Dhar, 2015; Ganguli et al., 2018). Recognising this limitation, the NPB-2018 has considerably expanded the types of allowable feedstocks for ethanol production under the blending programme and has a fixed guaranteed purchase price for ethanol made from these sources. The expanded list now includes sorghum, sugar beet, maize (corn), sugarcane juice, sugar, rotten potatoes, damaged food grains, and surplus rice lying with the Food Corporation of India (FCI).[1] Diversion of food grains for ethanol production is supposed to be based on the approval of the National Biofuel Coordination Committee (NBCC) (MoPNG, 2020; Sarwal et al., 2021). In fact, the Expert Committee constituted by NITI Aayog and MoPNG for developing the Ethanol Road Map for India 2020–2025 foresees ethanol production of about 10 billion litres from these alternative 1G sources required to achieve the 20 per cent blending target by 2025 (Sarwal et al., 2021). For comparison, this figure is more than three times the current ethanol production for blending (see Table 10.1), which mainly comes from sugarcane molasses; experts have labelled it "unrealistically optimistic" (O'Malley & Searle, 2021).

Increased reliance on 1G ethanol sources has significant unintended consequences that should not be disregarded by public policymakers. These include the following:

- It is prudent not to treat the jump in ethanol production in 2020–2021 as the new normal, given that this was possible due to a bumper crop. Sugarcane is a very water-intensive crop and bumper production is certainly not guaranteed in years of below-average rainfall (Chandra, 2020; O'Malley & Searle, 2021). If ethanol production from sugarcane molasses were to decrease in subsequent years, the pressure to increase production from other 1G feedstocks will increase in order to meet blending targets, which will result in downstream consequences.
- Due to favourable policies, it is already commercially attractive to supply ethanol for the biofuel blending programme. In the past few years, this has led to reduced ethanol supply to other industrial/potable uses, resulting in a sharp rise in ethanol imports by those industries, even though importing ethanol for the biofuels industry is prohibited (Chandra, 2020). This is an ironic outcome since one of the key goals of India's biofuels programme was to reduce the reliance on (petroleum) imports by which to save foreign exchange (O'Malley & Searle, 2021). The contradictory nature of importing alcohol instead of petroleum seems to be lost on policymakers!
- Increased demand and profitability of sugarcane as a consequence of the EBP is likely to lead to expansion of sugarcane cultivation. It would

be a serious mistake if India tried to emulate the "success" of Brazil's sugarcane-ethanol programme. Although population density is an admittedly crude measure of pressure on land, even a superficial comparison between India (455/km^2) and Brazil (25/km^2) demonstrates that the countries are leagues apart (World Bank, 2018). Expansion of sugarcane cultivation in India is likely to come at the expense of food crops since sugarcane needs good quality soil and abundant water (Beckman et al., 2018).

- Even the NITI Aayog-MoPNG Expert Committee openly admits that heavy reliance on sugarcane is problematic because sugarcane cultivation is water-intensive. Land under sugarcane cultivation is typically irrigated, but the increasing diversion of water for sugarcane automatically means less water for other crops and other users and ecological needs (O'Malley & Searle, 2021). Accordingly, the Expert Committee recommends the encouragement of less water-intensive alternatives for ethanol feedstock such as maize (corn), sorghum, and sugar beets (Sarwal et al., 2021). However, this gradual expansion of the use of food crops for 1G ethanol goes against the spirit of India's biofuels policy, in which non-competition with food crops is supposed to be a central pillar (O'Malley & Searle, 2021).
- The recent inclusion of damaged and surplus food grains as a potential ethanol feedstock, ostensibly with the approval of the NBCC, has been highly controversial in a food-insecure country (Hussain & Ranade, 2020; Agarwal, 2020; Hussain & Mohapatra, 2021). To critics, this appears like a "slippery slope" towards using food grains and therefore must be avoided completely.

10.4.2 Prospects for 2G Ethanol

Recognising the constraints of 1G ethanol, as well as the enormous theoretical potential of 2G ethanol, the GoI has tried to promote the latter in the NPB-2009 and more robustly through the NPB-2018. In addition to a range of financial incentives for 2G bio-refineries such as viability gap funding (VGF), 100 per cent foreign direct investment has been allowed and 15-year guaranteed purchase agreements have been promised (MoPNG, 2020; Gupta et al., 2020). Nearly INR 20 billion (about US$270 million) have been earmarked for the PM-JIVAN scheme focused on bio-refineries; extensive research is also being funded by the Department of Biotechnology, MoPNG, and the Ministry of New and Renewable Energy (Gupta et al., 2020). At present, there are only two 2G ethanol refineries at the pilot and demonstration stages in India. Their combined output is just less than two million litres per year. However, nearly a dozen new facilities are in the planning stages with a combined production capacity of more than 400 million litres per year (MoPNG, 2020).

Since 2G ethanol is derived from plant biomass, that is agricultural and forestry residues, it is important to estimate the availability of such residues for ethanol production. Many studies have attempted to calculate this available quantity since ethanol production cannot be estimated without a good grasp of feedstock availability. The Technology Information, Forecasting, and Assessment Council (TIFAC), a key think tank under the Ministry of Science and Technology, GoI, has estimated the total dry biomass generated in the agricultural sector at 683 million tonnes, out of which only 178 million tonnes were deemed to be "surplus" and thus potentially available for ethanol production. Based on this estimate, TIFAC calculated a 2G ethanol production potential of about 51 billion litres per year (TIFAC, 2018).

Many independent researchers and organisations have also tried to estimate 2G ethanol potential in India. Purohit and Chaturvedi (2018) have estimated crop residue production of 876 million tonnes by 2030, while Purohit and Dhar (2018) have come up with a very similar estimate of 887 million tonnes by 2030. However, the International Council on Clean Transportation (ICCT) points out that from the "gross" crop residue production estimates, it is important to deduct some amount for soil protection and soil carbon enrichment, as well as for already-existing uses such as livestock feed, livestock bedding, domestic fuel, and thatching for rural homes, to arrive at "net sustainable" supply of crop residues for ethanol production. Based on these corrections, the ICCT determines a quantity of 71 million tonnes of net sustainable crop residues by 2030, which translates into 26.7 billion litres on 2G ethanol (Pavlenko & Searle, 2019). It is important to recognise that this is an estimate of theoretical potential; even so, it is approximately half of the TIFAC's (2018) estimate. In reality, the ICCT study concludes that actual production is likely to be only a fraction of the theoretical potential due to supply chain, technological, and commercialisation challenges. Despite at least two decades of research and policy support, 2G ethanol production is still quite limited in the United States and EU, and it would be unrealistic to assume that India's experience would be dramatically different. Accounting for a range of challenges and trends, the ICCT study came up with a "realistic" production estimate of about two billion litres of 2G ethanol by 2030 (Pavlenko & Searle, 2019). For reference, this is approximately the amount of 1G ethanol currently being produced annually in India (see Table 10.1). A 2021 study by the ICCT further concludes that the current incentives offered by the government, centred around VGF, are insufficient to overcome the serious cost disadvantages faced by 2G ethanol producers, and much more significant policy support will be necessary for commercial viability (Zhou et al., 2021).

10.5 Biodiesel

As discussed in Section 10.3, India's original aim of promoting biodiesel production from tree-borne, non-edible oilseeds, such as Jatropha and Pongamia, through the NBM has effectively died a "quiet death" (Chandra, 2020), even though that may still be mentioned in government policy documents. Biodiesel producers have switched to other feedstocks to remain in operation, by far the most important of which is imported palm oil, accounting for about 70 per cent of current production (Chandra, 2020; MoPNG, 2020). The biggest exporters of palm oil are Indonesia and Malaysia, where the expansion of palm oil plantations has led to massive deforestation of old-growth rainforests with dire consequences for biodiversity and carbon sequestration (Vijay et al., 2016; IUCN, 2018; Lustgarten, 2018; Greenpeace, 2018; WWF, 2021). This effectively amounts to outsourcing of ecological harm for a domestic policy that is intended to benefit the environment. This contraction does not find a mention in any government policy document.

Biodiesel production in India has increased steadily over the past decade from about 100 million litres in 2010 to just over 220 million litres in 2020 (Chandra, 2020). Only half a dozen bio-refineries are operational and almost all are operating well below their rated capacity due to a combination of demand and feedstock supply constraints. About half of the biodiesel produced is used by a mix of local users such as backup power generators, farm equipment, small- and medium-sized industries like brick kilns, etc. About 40 per cent of biodiesel produced is blended for on-road use by the OMCs. Some prominent users are the Indian Railways, State Road Transport Corporations, port authorities, and so forth. The national average blending share has remained minuscule, increasing only slightly from 0.08 per cent in 2015 to 0.16 per cent in 2020 (Chandra, 2020). A small but growing share of biodiesel is exported, mainly to the EU, which is a lucrative export opportunity (Chandra, 2020). However, allowing exports appears to go against the policy goal of increasing the blending share in diesel to meet the 5 per cent target by 2030.

Since setting a more realistic blending target of 5 per cent (by 2030) in the NPB-2018, the GoI has started exploring the potential for alternative feedstocks seriously. UCO can be converted to biodiesel relatively easily; already about 25 per cent of biodiesel produced in India is from UCO (Chandra, 2020). About 27 billion litres of cooking oil is used annually in India, which represents a huge untapped potential. Moreover, the re-use of UCO, especially in small and medium food establishments, is a recognised public health challenge in India (FSSAI, 2018). In 2019, the Food Safety and Standards Authority of India issued a directive requiring food establishment operators to use more than 50 litres of cooking oil per day to maintain use and disposal records. Despite this positive step, observers feel that oversight and enforcement are unlikely to happen in the foreseeable future and big hotels and food processing companies will continue to remain the best source of UCO. Due to supply chain constraints, it has been estimated that only about 1.4 billion litres of UCO can be reliably collected in the near term to produce 1.1 billion litres of biodiesel (Chandra, 2020). Using less conservative assumptions about the collection, the ICCT estimates 3.3 billion litres of biofuel from UCO by 2030 (Pavlenko & Searle, 2019). In 2019, MoPNG established attractive procurement prices for biodiesel from UCO along with other incentives similar to those for ethanol mentioned in Section 10.4, and plans for six UCO-biodiesel plants have since been announced (Chandra, 2020).

Biodiesel can also be produced from animal fats and tallow. In fact, a small amount of such biodiesel is already being produced in India; however, there is no estimate of its potential in the literature. The GoI is also taking an active interest in "advanced" biofuel technologies such as synthetic "drop-in" biodiesel production from Municipal Solid Waste (MSW) using the Fischer-Tropsch process (Chandra, 2020). MSW management is a huge burden in India and the increasing generation of MSW in the coming decades is only going to make the problem more acute. Waste-to-energy plants are already proliferating, burning waste to produce electricity. Under these circumstances, MSW can appear attractive as a potential feedstock for "advanced" biodiesel, but such conversion technologies are at a relatively nascent stage and yet to be proven on a commercial scale (Chandra, 2020). The GoI has launched extensive research programmes and created incentives for pilot and demonstration plants. The ICCT estimates a theoretical potential for synthetic diesel from MSW at 13.4 billion litres, but due to the practical constraints mentioned earlier, the think tank projects actual production at slightly above one billion litres by 2030 (Pavlenko & Searle, 2019).

According to MoPNG (2020), diesel demand has been growing at approximately 3 per cent a year, ignoring the temporary impact of the COVID-19 pandemic. Extrapolating to 2030 produces a demand of 135 billion litres. Table 10.3 compares the 10.5 per cent blending target in 2030 with biodiesel production estimates from the literature summarised earlier.

To meet the 5 per cent blending target in 2030, a 30-times increase in production will be required compared to actual current biodiesel production volumes – in a matter of 10 years! This appears quite unlikely, unless the projected demand for diesel declines substantially. UCO-based biodiesel holds significant promise, having the potential to meet nearly half of the 5 per cent blending target in 2030. The MSW-based biodiesel projection is somewhat speculative since the technology is not yet commercially proven. If the GoI seriously pursues the 5 per cent blending target by 2030, it is likely that there will be a push to fill the "gap" with unsustainable feedstocks such as palm oil, leading to undesirable environmental outcomes.

Table 10.3 Comparing biodiesel demand and supply estimates for India in 2030.

Fuel	*Quantity (litres)*
Diesel demand in 2030	135 billion
Biodiesel demand at 5% blending in 2030	6.75 billion
UCO-biodiesel projection in 2030[a]	3.3 billion
MSW-biodiesel projection in 2030[a]	1.1 billion
Biodiesel actual production in 2020[b]	225 million

Source: [a] Pavlenko & Searle (2019); [b] Chandra (2020)

10.6 Conclusion: The Need for Holistic Thinking About Transport Sector Sustainability

The preceding sections have demonstrated that scaling up biofuels sustainably would be extremely challenging for the foreseeable future in India. The following three points summarise the analysis thus far:

(1) The 1G ethanol production can be increased substantially only if socio-economic and environmental considerations are disregarded, especially the food-versus-fuel conflict. Industrial ethanol users are already facing a supply crunch leading to increased ethanol imports, which defeats the purpose of saving foreign exchange by reducing petroleum imports. Even with increased 1G ethanol production, meeting the 20 per cent ethanol blending target may prove to be elusive if petrol demand continues to grow at predicted rates.
(2) The 2G ethanol has not made a substantial contribution globally despite enormous efforts, due to technological and commercial challenges. It is unrealistic to assume that it will contribute significantly to India's 2030 ethanol blending targets, even if theoretical studies demonstrate the existence of ample feedstocks.
(3) Most biodiesel is currently made from imported palm oil feedstock, which causes ecological disaster elsewhere and therefore cannot be said to contribute towards sustainability goals. UCO holds high promise as a sustainable solution and has the potential to contribute significantly towards biodiesel blending targets. Biodiesel from MSW is still not commercially proven, though research and development should continue.

Against this backdrop, a simple switch from conventional oil fuels to biofuels is far more complicated than envisaged. It behoves us to think more holistically about the entire transport sector. Even a cursory examination of the transport sector in India throws up some obvious and uncomfortable realities. Table 10.4 depicts the energy consumption share for transport by mode and fuel respectively in India.

Table 10.4 shows that India is heavily reliant on the road sector, which is almost exclusively dependent on petroleum, at least at present. In contrast,

Table 10.4 Energy consumption for transport by mode in India, 2017.

Transport mode	*Share of energy consumption (%)*
Road	90
Rail	4
Domestic aviation	4
Other modes*	2

Note: * Other modes include waterways and pipeline transport.

Source: IEA (2020)

rail, already highly electrified and easily amenable to further electrification, is only 4 per cent of the modal energy share. Moreover, rail is many times more energy-efficient than road-based transport. According to the IEA (2019a), passenger rail is four to ten times more energy-efficient per passenger-km (depending on the type of road transport), and freight rail is seven to ten times more energy-efficient per tonne-km (depending on the type of road-based freight trucks). This means that a relative modal shift from road to rail, even when the rail is not electrified, would result in significant savings in petroleum fuels. Therefore, any reasonable strategy to reduce petroleum dependence, whether to reduce import dependence or to address environmental concerns, ought to focus heavily on reducing road transport dependence.

Unfortunately, while India is among the top few countries in the world in terms of the rail network, and rail transport has indeed been increasing steadily over recent decades, the share of rail in both passenger transport (in terms of passenger-km) and freight transport (in terms of tonne-km) has been falling compared to other transport modes (IEA, 2020). This means that the boom in road transport is far surpassing rail expansion. Over the past 30 years, road freight transport in India has increased by a factor of ten, and its associated oil demand has more than tripled since 2000, making it the highest growth rate in the world (IEA, 2020). In terms of passenger vehicles, India is already the world's fourth-largest market and is expected to become the third-largest within a few years. Although two-wheelers still dominate the passenger vehicle segment, there has been a worrying rise in fuel-inefficient SUVs in recent years. The GoI is actively pushing growth in the automotive sector with an aspirational goal of making India one of the leading automotive manufacturing hubs in Asia. The sector currently contributes about 7 per cent to India's GDP and the government's Automotive Mission Plan 2016–2026 hopes to increase this to 12 per cent (Gupta et al., 2018). If current trends continue, the IEA (2021) forecasts a doubling of energy demand for road transport by 2040. All this clearly implies that Indian policymakers do not appear to see any contradiction between the continued growth of road-based transport and environmental sustainability concerns.

A detailed analysis of overall sustainability strategies for the transportation sector in India is outside the scope of this chapter. Still, a few obvious directions emerge naturally from the preceding discussion.

(1) *Widespread and rapid electrification*: India's electricity generation is still highly coal-dependent but renewables are scaling up rapidly, and the transition is widely expected to accelerate significantly in the years ahead (IEA, 2021). Electric vehicles (EVs), in addition to providing urban air quality benefits, also produce less life cycle GHG emissions, even when electricity is generated from fossil fuels since electric drivetrains are much more efficient than internal combustion engine vehicles (DoE, n.d.; Bieker, 2021). Therefore, rapid electrification of transportation has the potential to significantly reduce the demand for petroleum fuels as well

as GHG emissions. Candidates for rapid electrification in road transport include two- and three-wheelers, cars, urban buses, and light commercial vehicles. Two- and three-wheelers together consume more than 60 per cent of petrol in India (Nielsen, 2013). These types of vehicles are amenable to electrification relatively easily and quickly, compared to cars, trucks, and buses. In China, already nearly 70 per cent of new two-wheeler sales are electric (McKerracher, 2021). The rapid uptake of electric two- and three-wheelers could achieve petrol substitution at a much faster and larger scale than problematic bio-ethanol blending targets. Unfortunately, while India has adopted transport electrification strategies, analysts believe that unless bolstered by mandates, as witnessed in China (He & Jin, 2021), adoption of EVs is likely to remain painfully slow (Rokadiya & Callahan, 2019; Goel et al., 2021; Bandivadekar, 2021), and insufficient to make a noticeable dent in petroleum demand in the foreseeable future.

(2) *Modal shift to rail*: Rail is among the most efficient forms of transport but, as noted in the preceding paragraph, India is moving steadily in the wrong direction, namely increasing share of road transport. There is an urgent need for holistic, multisectoral planning and a policy push to reverse this dangerous trend. This is especially true for freight transport, in which India has one of the highest growth rates in the world (IEA, 2020). Rail electrification is proceeding at a commendable rate (CORE, 2021) but maybe further speeded up if deemed a priority.

(3) *Reducing automobile dependence*: India's rapidly growing middle class has been fuelling spectacular growth in car and two-wheeler sales over the past two decades, due to a combination of necessity and "status." The government has played an actively encouraging role, given the automotive sector's large contributions to GDP and employment (Gupta et al., 2018). India lags severely in providing appropriate mass transit options, especially outside limited zones in a handful of metro cities (Baindur, 2015; Abhishek, 2020). In recent decades, short-sighted land-use planning has led to the creation of numerous satellite townships and employment hubs in suburban and ex-urban areas where mass transit options are virtually non-existent (Mishra, 2019). Unless these trends are reversed, reducing reliance on petroleum fuels will be very challenging.

(4) *Increasing vehicle fuel economy*: India introduced fuel-efficiency standards for passenger cars for the first time in 2015, much later than in many other countries. Even so, the 2022 standards are nearly 20 per cent less stringent than those in the EU (Bandivadekar, 2021). Although fuel-efficiency standards for heavy-duty vehicles were introduced recently, two-wheelers, which are responsible for nearly two-thirds of petrol consumption in India, are not covered by any standards (Bandivadekar, 2021). Expansion and tightening of standards can make a significant dent in overall transportation fuel use as well as save consumers money, as the experience in other countries clearly demonstrates (IEA, 2019b).

It should be obvious that blindly allowing ever-increasing growth in demand for petroleum fuels and then trying to substitute a part of it with biofuels is an indefensible strategy. If India is serious about avoiding catastrophic climate change, reducing demand for petroleum fuels, using the approaches mentioned earlier, ought to become a national priority, at par with scaling up renewable electricity. Substantial co-benefits will accrue from reducing petroleum fuel consumption in terms of lowered import bills and public health. A significant demand reduction in petroleum fuels will make biofuel blending targets easier to meet while also "buying time" to come up with alternative solutions for the difficult-to-substitute sectors such as heavy commercial vehicles, aviation, and maritime transport. It is too early to predict which green technologies, whether second- or third-generation biofuels, green hydrogen, or something else, will become viable for these transport sectors in the future; research and development efforts must continue in earnest. But pursuing misguided blending targets, while ignoring trade-offs and demand-side measures, will do little to promote environmental sustainability.

Note

1 The FCI procures food grains directly from farmers at a guaranteed price and makes them available for the public distribution system, thus playing a major role in farmer security and food security.

References

Abhishek, V. (2020). Inadequate urban transportation facilities leave the poor in India high and dry. *Economic and Political Weekly*, 55(28–29). Retrieved from: www.epw.in/engage/article/inadequate-urban-transportation-facilities-leave (Last accessed: 25.07.2021)

Agarwal, K. (2020). As millions fear starvation, govt to convert 'surplus' rice into ethanol for sanitiser. *The Wire*. 21.04.2020. Retrieved from: https://thewire.in/government/pds-rice-ethanol-sanitiser (Last accessed: 15.06.2021)

Aradhey, A. (2019). *Biofuels Annual 2019: India. Global Agricultural Information Network (GAIN) Report #IN9069*. US Department of Agriculture Foreign Agricultural Service. Retrieved from: https://apps.fas.usda.gov/newgainapi/api/report/downloadreportbyfilename?filename=Biofuels%20Annual_New%20Delhi_India_8-9-2019.pdf (Last accessed: 05.06.2021)

Ariza-Montobbio, P. & Lele, S. (2010). Jatropha plantations for biodiesel in Tamil Nadu, India: Viability, livelihood trade-offs, and latent conflict. *Energy Policy*, 70(2): 189–195.

Baindur, D. (2015). *Urban Transport in India: Challenges and Recommendations*. Indian Institute of Human Settlements. Retrieved from: https://iihs.co.in/knowledge-gateway/wp-content/uploads/2015/07/RF-Working-Paper-Transport_edited_09062015_Final_reduced-size.pdf (Last accessed: 25.07.2021)

Bandivadekar, A. (2021). Being petroleum-independent. *The Hindu*. 24.02.2021. Retrieved from: www.thehindu.com/opinion/op-ed/being-petroleum-independent/article33916640.ece (Last accessed: 25.09.2021)

Beckman, J., Gooch, E., Gopinath, M. & Landes, M. (2018). Market impacts of China and India meeting biofuel targets using traditional feedstocks. *Biomass and Bioenergy*, 108(2018): 258–264.

Bieker, G. (2021). *A Global Comparison of Life-Cycle Greenhouse Gas Emissions of Combustion Engine and Electric Passenger Cars*. The International Council on Clean Transportation (ICCT). Retrieved from: https://theicct.org/publications/global-LCA-passenger-cars-jul2021 (Last accessed: 17.09.2021)

Biswas, P.K., Pohit, S. & Kumar, R. (2010). Biodiesel from jatropha: Can India meet the 20% blending target? *Energy Policy*, 38(2010): 1477–1484.

BP (2019). *BP Energy Outlook 2019 Edition*. British Petroleum. Retrieved from: www.bp.com/content/dam/bp/business-sites/en/global/corporate/pdfs/energy-economics/energy-outlook/bp-energy-outlook-2019.pdf (Last accessed: 23.04.2021)

Bracmort, K. (2020). *The Renewable Fuel Standard (RFS): An Overview. CRS Report #R43325*. Congressional Research Service, Washington, DC. Retrieved from: www.everycrsreport.com/files/20200414_R43325_1981cda6b9497b16f7306b81d584cbcf91d4c801.pdf (Last accessed: 10.06.2021)

Broom, D. (2020). *6 of the World's 10 Most Polluted Cities are in India*. World Economic Forum. Retrieved from: www.weforum.org/agenda/2020/03/6-of-the-worlds-10-most-polluted-cities-are-in-india/ (Last accessed: 21.04.2021)

CEIC (2023). *India Crude Oil: Imports*. CEIC Data, an ISI Emerging Markets Group Company, London, UK. Retrieved from: https://www.ceicdata.com/en/indicator/india/crude-oil-imports (Last accessed 25.01.2024)

Chandra, A. (2020). *Biofuels Annual 2020: India. Global Agricultural Information Network (GAIN) Report #IN2020-0122*. US Department of Agriculture Foreign Agricultural Service. Retrieved from: www.fas.usda.gov/data/india-biofuels-annual-5 (Last accessed: 10.06.2021)

CORE (2021). *Ministry of Railways, Government of India*. Central Organization for Railway Electrification. Retrieved from: https://core.indianrailways.gov.in/ (Last accessed: 05.08.2021)

Das, S. (2020). The national policy of biofuels in India – a perspective. *Energy Policy*, 143(2020): 111595.

De Oliveira, M.E., Vaughan, B.E. & Rykiel, E.J. (2005). Ethanol as fuel: Energy, carbon dioxide balances, and ecological footprint. *Bioscience*, 55(7): 593–602.

DoE (n.d.). *Reducing Pollution with Electric Vehicles*. Office of Energy Efficiency and Renewable Energy, U.S. Department of Energy. Retrieved from: www.energy.gov/eere/electricvehicles/reducing-pollution-electric-vehicles (Last accessed: 03.08.2021)

EC (2020). *Biofuels*. European Commission. Retrieved from: https://ec.europa.eu/energy/topics/renewable-energy/biofuels/overview_en (Last accessed: 23.04.2021)

EIA (2019). *International Data on Biofuels*. U.S. Energy Information Administration. Retrieved from: www.eia.gov/international/data/world/biofuels/biofuels-production (Last accessed: 12.05.2021)

EIA (2020). *Biofuels Explained*. U.S. Energy Information Administration. Retrieved from: www.eia.gov/energyexplained/biofuels/use-of-biodiesel.php (Last accessed: 18.06.2021)

EPA (2018). *Biofuels and the Environment: Second Triennial Report to Congress, EPA/600/R-18/195*. U.S. Environmental Protection Agency, Washington, DC. Retrieved from: https://cfpub.epa.gov/si/si_public_record_report.cfm?Lab=IO&dirEntryId=341491 (Last accessed: 20.06.2021)

EPA (2021). *Overview for Renewable Fuel Standard*. U.S. Environmental Protection Agency. Retrieved from: www.epa.gov/renewable-fuel-standard-program/overview-renewable-fuel-standard (Last accessed: 03.07.2021)

FAO (2013). *Biofuels and Food Security. A Report by the High Level Panel of Experts on Food Security and Nutrition of the Committee on World Food Security*. Food and Agriculture Organization of the United Nations, Rome.

Findlater, K.M. & Kandlikar, M. (2011). Land use and second-generation biofuel feedstocks: The unconsidered impacts of Jatropha biodiesel in Rajasthan, India. *Energy Policy*, 39(6): 3404–3413.

Flynn, K. (2017). Algal biofuel production is neither environmentally nor commercially sustainable. *The Conversation*. 08.08.2017. Retrieved from: https://theconversation.com/algal-biofuel-production-is-neither-environmentally-nor-commercially-sustainable-82095 (Last accessed: 19.07.2021)

FSSAI (2018). *Background Note on Used Cooking Oil*. Food Safety and Standards Authority of India. Retrieved from: https://fssai.gov.in/ruco/file/Background%20Note-UCO.pdf (Last accessed: 01.08.2021)

Ganguli, S., Somani, A., Motkuri, R.K. & Bloyd, C. (2018). *India Alternative Fuel Infrastructure: The Potential for Second-Generation Biofuel Technology*. Pacific Northwest National Laboratory (PNNL-28283), US Department of Energy. Retrieved from: www.osti.gov/servlets/purl/1530891/ (Last accessed: 10.07.2021)

Goel, P., Sharma, N., Mathiyazhagan, K. & Vimal, K. (2021). Government is trying but consumers are not buying: A barrier analysis for electric vehicle sales in India. *Sustainable Production and Consumption*, 28: 71–90.

Goldemberg, J., Coelho, S.T. & Guardabassi, P. (2008). The sustainability of ethanol production from sugarcane. *Energy Policy*, 36(6): 2086–2097.

Greenpeace (2018). *Final Countdown: Now or Never to Reform the Palm Oil Industry*. Greenpeace International, Amsterdam. Retrieved from: www.greenpeace.org/international/publication/18455/the-final-countdown-forests-indonesia-palm-oil/ (Last accessed: 01.08.2021)

Gupta, R., Suresh, D., Puri, K. & Ramakumar, S. (2020). India biofuel economy roadmap – challenges and prospects. In: *Commercializing Conventional and Advanced Transport Biofuels from Biomass and Other Renewable Feedstocks. Task 39 Newsletter Issue #56*. IEA Bioenergy. December 2020. Retrieved from: http://task39.sites.olt.ubc.ca/files/2020/12/IEA-Bioenergy-Task-39-Newsletter-Issue-56-Final-Draft.pdf (Last accessed: 15.06.2021)

Gupta, S., Huddar, N., Iyer, B. & Moller, T. (2018). *The Future of Mobility in India's Passenger-Vehicle Market*. McKinsey & Co. Retrieved from: www.mckinsey.com/industries/automotive-and-assembly/our-insights/the-future-of-mobility-in-indias-passenger-vehicle-market (Last accessed: 20.07.2021)

He, H. & Jin, L. (2021). *How China Put Nearly 5 Million New Energy Vehicles on the Road in One Decade*. The International Council on Clean Transportation (ICCT). Retrieved from: https://theicct.org/blog/staff/china-new-energy-vehicles-jan2021 (Last accessed: 20.08.2021)

Hussain, S. & Mohapatra, J. (2021). India wants to use food grain stock for ethanol. That's a problem in a hungry country. *The Print*. 01.07.2021. Retrieved from: https://theprint.in/opinion/india-wants-to-use-foodgrain-for-ethanol-thats-a-problem-in-a-hungry-country/686707/ (Last accessed: 20.09.2021)

Hussain, S. & Ranade, A. (2020). Diverting rice to produce ethanol during pandemic is unethical. *The Indian Express*. 06.05.2020. Retrieved from: https://indianexpress.com/article/opinion/columns/rice-to-produce-ethanol-hand-sanitisers-coronavirus-6395644/ (Last accessed: 15.06.2021)

IEA (2019a). *The Future of Rail: Opportunities for Energy and the Environment*. International Energy Agency. Retrieved from: www.iea.org/reports/the-future-of-rail (Last accessed: 15.08.2021)

IEA (2019b). *Fuel Economy in Major Car Markets: Technology and Policy Drivers 2005–2017*. International Energy Agency. Retrieved from: www.iea.org/reports/fuel-economy-in-major-car-markets (Last accessed: 25.09.2021)

IEA (2020). *India 2020: Energy Policy Review*. International Energy Agency. Retrieved from: www.iea.org/reports/india-2020 (Last accessed: 10.08.2021)

IEA (2021). *India Energy Outlook 2021*. International Energy Agency. Retrieved from: www.iea.org/reports/india-energy-outlook-2021 (Last accessed: 22.05.2021)

IRENA (2019). *Advanced Biofuels. What Holds them Back?* International Renewable Energy Agency, Abu Dhabi. Retrieved from: https://irena.org/-/media/Files/

IRENA/Agency/Publication/2019/Nov/IRENA_Advanced-biofuels_2019.pdf (Last accessed: 01.06.2021)

IUCN (2018). *Palm Oil and Biodiversity*. International Union for Conservation of Nature, Issues Brief. June 2018. Retrieved from: www.iucn.org/sites/dev/files/iucn_issues_brief_palm_oil_and_biodiversity.pdf (Last accessed: 01.08.2021)

Jeswani, H.K., Chilvers, A. & Azapagic, A. (2020). Environmental sustainability of biofuels: A review. *Proceedings of the Royal Society A*, 476: 20200351. http://doi.org/10.1098/rspa.2020.0351 (Last accessed: 05.06.2021)

Kant, P. & Wu, S. (2011). The extraordinary collapse of jatropha as a global biofuel. *Environmental Science & Technology*, 45(17): 7114–7115.

Kumar, S., Chaube, A. & Jain, S.K. (2012). Critical review of jatropha biodiesel promotion policies in India. *Energy Policy*, 41(2012): 775–781.

Lustgarten, A. (2018). Palm oil was supposed to help save the planet. Instead it unleashed a catastrophe. *The New York Times*. 20.11.2018. Retrieved from: www.nytimes.com/2018/11/20/magazine/palm-oil-borneo-climate-catastrophe.html (Last accessed: 24.07.2021)

McKerracher, C. (2021). The EV revolution rides on two wheels. *Bloomberg*. Retrieved from: www.bloomberg.com/news/newsletters/2021-04-20/hyperdrive-daily-the-ev-revolution-rides-on-two-wheels (Last accessed: 20.08.2021)

Ministry of Railways (2003). *Indian Railways to Use Bio-Diesel Fuel: Signs MoU with IOC for Production of Bio-Diesel*. Press Information Bureau, Government of India. Retrieved from: https://archive.pib.gov.in/archive/releases98/lyr2003/rfeb2003/12022003/r120220033.html (Last accessed: 01.07.2021)

Mishra, A. (2019). Cities, transport and agglomeration: Addressing the urban mobility challenges in India. *Growth and Change*, 50(3): 1115–1133.

MoEFCC (2021). *India: Third Biennial Update Report to the United Nations Framework Convention on Climate Change*. Ministry of Environment, Forest and Climate Change, Government of India. Retrieved from: https://unfccc.int/sites/default/files/resource/INDIA_%20BUR-3_20.02.2021_High.pdf (Last accessed: 04.04.2021)

MoPNG (2020). *Ministry of Petroleum and Natural Gas Annual Report 2019–20*. Government of India, New Delhi.

Mosnier, A., Havlik, P., Valin, H., Baker, J.S., Murray, B.C., Feng, S., Obersteiner, M., McCarl, B.A., Rose, S.K. & Schneider, U.A. (2012). *The Net Global Effects of Alternative U.S. Biofuel Mandates: Fossil Fuel Displacement, Indirect Land Use Change, and the Role of Agricultural Productivity Growth. Report NI R 12-01*. Nicholas Institute for Environmental Policy Solutions, Duke University, Durham, North Carolina. Retrieved from: https://nicholasinstitute.duke.edu/sites/default/files/publications/net-global-effects-of-alternative-u.s.-biofuel-mandates-paper.pdf (Last accessed: 01.07.2021)

Nielsen (2013). *All India Study on Sectoral Demand of Diesel and Petrol. Report Prepared for the Petroleum Planning & Analysis Cell*. Ministry of Petroleum & Natural Gas, Government of India. Retrieved from: www.ppac.gov.in/WriteReadData/Reports/201411110329450069740AllIndiaStudyonSectoralDemandofDiesel.pdf (Last accessed: 13.07.2021)

NRC (2011). *Renewable Fuel Standard: Potential Economic and Environmental Effects of U.S. Biofuel Policy*. National Research Council. National Academies Press, Washington, DC. Retrieved from: www.nationalacademies.org/our-work/economic-and-environmental-impacts-of-increasing-biofuels-production (Last accessed: 28.06.2021)

Nuelle, B. (2019). Next-gen biofuel dreams fade; developers blame EPA. *Agri-Pulse*. 11.12.2019. Retrieved from: www.agri-pulse.com/articles/12894-cellulosic-ethanol-struggles-to-climb-commercialization-ladder (Last accessed: 17.07.2021)

O'Malley, J. & Searle, S. (2021). *India, Don't Fall for Ethanol: Roadmap Leads National Policy on Biofuels Off Track*. The International Council on Clean

Transportation (ICCT). Retrieved from: https://theicct.org/blog/staff/india-ethanol-biofuels-policy-aug21 (Last accessed: 20.09.2021)

Oxfam (2008). *Another Inconvenient Truth: How Biofuel Policies are Deepening Poverty and Accelerating Climate Change. Oxfam Briefing Paper #114.* Oxfam International, Oxford, UK. Retrieved from: www.foe.ie/assets/files/pdf/oxfam_report_on_biofuels_june_2008.pdf (Last accessed: 27.06.2021)

Padella, M., O'Connell, A. & Prussi, M. (2019). What is still limiting the deployment of cellulosic ethanol? Analysis of the current status of the sector. *Applied Sciences*, 9: 4523. https://doi.org/10.3390/app9214523

Panoutsou, C., Germer, S., Karka, P., Papadokostantakis, S., Kroyan, Y., Wojcieszyk, M., Maniatis, K., Marchand, P. & Landalv, I. (2021). Advanced biofuels to decarbonise European transport by 2030: Markets, challenges, and policies that impact their successful market uptake. *Energy Strategy Reviews*, 34(2021): 100633.

Pavlenko, N. & Searle, S. (2019). *The Potential for Advanced Biofuels in India: Assessing the Availability of Feedstocks and Deployable Technologies, Working Paper 2019–22*. The International Council on Clean Transportation (ICCT). Retrieved from: https://theicct.org/publications/potential-advanced-biofuels-india-assessing-availability-feedstocks-and-deployable (Last accessed: 14.07.2021)

Purohit, P. & Chaturvedi, V. (2018). Biomass pellets for power generation in India: A techno-economic evaluation. *Environmental Science and Pollution Research*, 25(29): 29614–29632.

Purohit, P. & Dhar, S. (2015). *Biofuel Roadmap for India*. UNEP (United Nations Environment Programme) and DTU (Denmark Technical University) Partnership. Retrieved from: https://unepdtu.org/publications/biofuel-roadmap-for-india/ (Last accessed: 05.07.2021)

Purohit, P. & Dhar, S. (2018). Lignocellulosic biofuels in India: Current perspectives, potential issues and future prospects. *AIMS Energy*, 6(3): 453–486.

Renzaho, A., Kamara, J.K. & Toole, M. (2017). Biofuel production and its impact on food security in low and middle income countries: Implications for the post-2015 sustainable development goals. *Renewable and Sustainable Energy Reviews*, 78: 503–516.

Rokadiya, S. & Callahan, J. (2019). *Boosting EV Growth in India: Missing Pieces in the Policy Puzzle*. The International Council on Clean Transportation. Retrieved from: https://theicct.org/blog/staff/boosting-ev-growth-india-20191028 (Last accessed: 20.08.2021)

Sarwal, R., Kumar, S., Mehta, A., Varadan, A., Singh, S.K., Ramakumar, S.S.V. & Mathai, R. (2021). *Roadmap for Ethanol Blending in India 2020–2025: Report of the Expert Committee*. NITI Aayog and Ministry of Petroleum & Natural Gas, New Delhi.

Stattman, S., Gupta, A., Partzsch, L. & Oosterveer, P. (2018). Toward sustainable biofuels in the European Union? Lessons from a decade of hybrid biofuel governance. *Sustainability*, 10: 4111. https://doi.org/10.3390/su10114111

Subramaniam, Y., Masron, T.A. & Azman, N.H. (2019). The impact of biofuels on food security. *International Economics*, 160: 72–83.

T&E (2021). *10 Years of EU Fuels Policy Increased EU's Reliance on Unsustainable Biofuels. Transport & Environment Briefing Paper*. Retrieved from: www.transportenvironment.org/sites/te/files/publications/Biofuels%20briefing%20072021.pdf (Last accessed: 11.08.2021)

Tenenbaum, D.J. (2008). Food vs. fuel: Diversion of crops could cause more hunger. *Environmental Health Perspectives*, 116(6): A254–A257.

TIFAC (2018). *Estimation of Surplus Crop Residues in India for Biofuel Production*. Technology Information, Forecasting and Assessment Council, Ministry of Science and Technology, Government of India, New Delhi.

TP (n.d.). *Brazil: Biofuels*. A Collaboration between The International Council on Clean Transportation (ICCT) and DieselNet. Retrieved from: www.transportpolicy.net/standard/brazil-fuels-biofuels/ (Last accessed: 03.05.2021)

Valin, H., Peters, D., van den Berg, M., Frank, S., Havlik, P., Forsell, N. & Hamelinck, C. (2015). *The Land Use Change Impact of Biofuels Consumed in the EU: Quantification of Area and Greenhouse Gas Impacts*. International Institute for Applied Systems Analysis (IIASA). Retrieved from: https://ec.europa.eu/energy/sites/ener/files/documents/Final%20Report_GLOBIOM_publication.pdf (Last accessed: 16.06.2021)

Vijay, V., Pimm, S.L., Jenkins, C.N. & Smith, S.J. (2016). The impact of oil palm on recent deforestation and biodiversity loss. *PLoS One*, 11(7): e0159668. https://doi.org/10.1371/journal.pone.0159668

World Bank (2018). *Population Density Data by Country*. Retrieved from: https://data.worldbank.org/indicator/EN.POP.DNST (Last accessed: 21.06.2021)

WWF (2021). *Deforestation Fronts: Drivers and Responses in a Changing World*. World Wide Fund for Nature, Gland, Switzerland. Retrieved from: https://wwfint.awsassets.panda.org/downloads/deforestation_fronts___drivers_and_responses_in_a_changing_world___full_report_1.pdf (Last accessed: 02.07.2021)

Zhou, Y., Searle, S. & Anup, S. (2021). *Techno-Economic Analysis of Cellulosic Ethanol in India Using Agricultural Residues, White Paper June 2021*. The International Council on Clean Transportation (ICCT). Retrieved from: https://theicct.org/sites/default/files/publications/cellulosic-ethanol-analysis-india-jun2021.pdf (Last accessed: 20.09.2021)

11 Business Model Innovations for Electric Vehicle Adoption in India

An Ecosystem Perspective

Annika Bose Styczynski

11.1 India's Electric Vehicle Sector: An Overview

The Indian electric vehicle (EV) market has grown significantly since January 2021. The initiation of the Faster Adoption and Manufacturing of (Hybrid and) Electric Vehicles (FAME) programme of the Government of India in 2015 has resulted in a marked increase in domestic production and sales of EVs, with the largest growth being observed in the electric two- and three-wheeler segments. Despite a steady growth in EV sales, the goal of the National Electric Mobility Mission Plan (NEMMP) of putting 6–7 million EVs on India's roads by 2020 proved difficult to attain. The officially registered stock of more than 576,600 EVs at the end of 2020, even when factoring in the large hidden figure of low-speed electric two-wheelers and e-rickshaws that are exempt from registration, was not anywhere near the envisioned volumes. However, with a stock of more than 2.5 million EVs in September 2023, India is an important emerging market of electric mobility (Vahan Sewa Dashboard, 2023). The scale of India's EV utilisation will ultimately have a more significant impact on aggregated greenhouse gas emissions compared to the scale of utilisation in smaller markets (Bose Styczynski & Hughes, 2019). At the end of March 2021, the monthly EV registration figures started to exceed pre-COVID-19 levels, which signify the breakthrough recovery of this sector (see Figure 11.1).

Towards the end of 2020, EVs constituted about 1 per cent of the official annual vehicle sales, which marks the threshold value between a niche and an early mass market. In September 2023, the share of EVs in total vehicle registrations reached 6.8 per cent (own calculation based on Vahan Sewa Dashboard figures). Given the dampened vehicle sales since fiscal year 2019–2020, this shows that EVs have started to have a displacement effect.

There is credible interest in the government to lead the Electric Mobility Mission to success, which is underpinned by at least three major motives:

(1) Energy security
India covered 88 per cent of its oil needs through imports in the fiscal year 2018–2019. The transport sector is the largest consumer, accounting

DOI: 10.4324/9781003281818-11

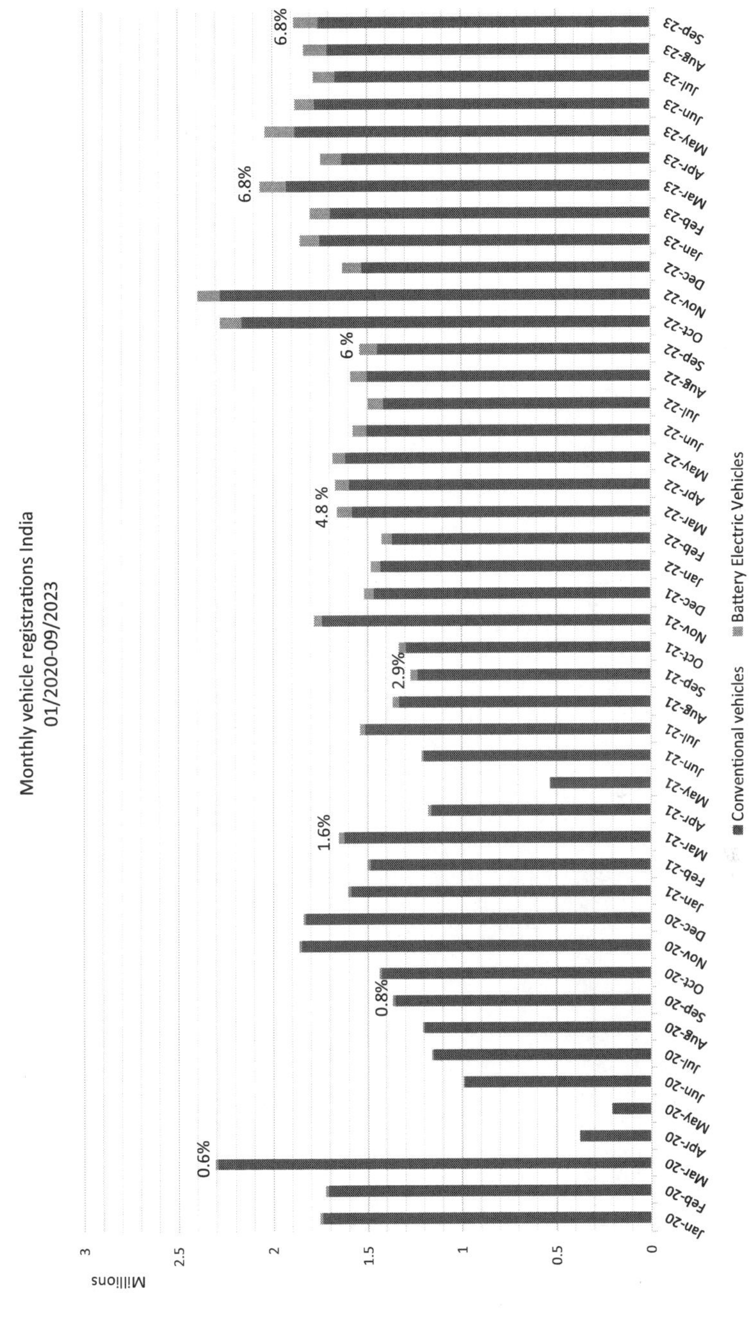

Figure 11.1 Monthly EV sales in India 2020–2023.

Source: Vahan Sewa Dashboard

for one-third of India's oil demand. Within this sector, road transport consumes about 80 per cent of the total demand (GIZ, 2021). A shift to electric mobility can considerably reduce the national economy's dependency on oil imports and the related price and exchange rate volatility.

(2) Domestic EV manufacturing
Two-wheelers dominate about 80 per cent of the Indian vehicle market (Society of Indian Automobile Manufacturers [SIAM]). A thriving domestic EV production can reduce costs while taking into consideration various local factors such as mineral resource availability and other battery production capabilities.

(3) Mitigating local air pollution and climate change
Vehicular exhausts are a major contributor to air pollution in urban India. Petrol and diesel are still the two major vehicle fuels (Vahan Sewa Dashboard, 2023). Responsible for 8.6 per cent of the total CO_2-eq emissions, the transport sector is the fourth-largest source of greenhouse gases in the country (Ge & Friedrich, 2020). Powered by renewable sources of energy, battery-operated vehicles can make a significant contribution to lowering local air pollution and mitigating climate change (Bieker, 2021).

Phase I of the FAME programme began in 2015, covering two fiscal years initially. However, it was extended twice until March 2019. The total funding outlay was INR 895 crores (US$120 million). Like its successor scheme, FAME prioritised demand creation. The Government of India (GoI, 2012) argues that the "vicious circle" of high manufacturing costs and low demand must be broken by creating a conducive environment that assures the industry of the following:

- Specific continuous minimum demand volumes.
- Commitment and support from the government especially in the nascent phases of this industry.
- Stability of government policies through an agreed-upon medium- to long-term plan.

This first phase of the funding scheme also announced public investments in the range of about 20–25 per cent of the overall funding volume for Technology Platform and Pilot Projects. However, FAME-I failed in terms of overall fund utilisation. In fact, only INR 359 crore (40 per cent) of the total funding was spent (GIZ, 2021). It remains largely unclear how this amount was used for indigenous product development and to overcome related challenges.

Looking ahead, NITI Aayog and the Rocky Mountains Institute (RMI) have developed two scenarios for India's EV market in 2030. Given the lack of an official government target, these scenarios provide a framework for orientation (NITI Aayog and RMI, 2017) as presented in Table 11.1.[1] In the business as usual (BAU) scenario, EVs across different modes have less than

Table 11.1 EV market share in India by 2030.

EV type	*Fiscal year (2019–2020)*	*FAME phase II (2019–2022)*	*NITI Aayog and RMI (2030)*	
Total no. of registrations	246,000		BAU (%)	TSF (%)
Two-wheeler	152,000	1 million	5	40
Three-wheeler	90,000	0.5 million	5	100
Four-wheeler	3,400			
Passenger cars			1	40
Commercial vehicles		55,000	5	100
Public transport			1	100
Buses	600	7,090		
Etc.				

Source: Authors' own.

5 per cent market share. In the transformative scenario (TSF), electric two-wheelers and passenger cars have a 40 per cent market share, while electric three-wheelers, commercially used four-wheelers, and public transit are completely electrified (ibid.).

Phase II of the FAME programme commenced in the FY2019–2020. It largely resonates with NITI's mode-based distinctions, specifying government funding priorities along similar lines. With FAME-II (2019–2022), the Central Government would put "greater emphasis on providing affordable and environmentally friendly public transportation options for the masses." The scheme would be "applicable mainly to vehicles used for public transport or those registered for commercial purposes in 3W, 4W, and bus segments. However, privately owned registered 2Ws will also be covered under the scheme as a mass segment" (The Gazette of India, 2019). In 2019, the Ministry of Finance reduced the Goods and Services Tax (GST) rate for the purchase of EVs from 12 per cent to 5 per cent and announced an income tax rebate of INR 1.5 lakh (US$2,000) on the purchase of an EV to support the scheme. Internal combustion engine (ICE) vehicles, by contrast, are subject to a GST rate of up to 28 per cent, which shows that Indian Government is applying a bonus-malus regulation to steer the market.

The funding for FAME-II amounts to INR 10,000 crores (US$1.4 billion), which includes the INR 366 crores spillover from FAME-I to be disbursed in tranches over a three-year period. The scheme covers the financial support for EV charging infrastructure and administrative expenditures; however, about 86 per cent of the outlay is allocated for spurring the demand of up to 1.56 million vehicles across segments (GIZ, 2021) as outlined in Table 11.1.

Along with FAME-II, the Cabinet approved the National Mission on Transformative Mobility and Battery Storage, an initiative by NITI Aayog, dedicated to the creation of a Phased Manufacturing Programme (PMP) for EVs, EV components, and batteries. Over a period of five years, initially, the PMP will focus on domestic manufacturing and assembly, promoting localised production across the entire EV value chain including large-scale, export-competitive battery and cell-manufacturing. Taking these ambitions into consideration, a holistic EV value chain consists of a number of segments, including (1) raw material extraction; (2) material processing; (3) battery, vehicle, and supply equipment manufacturing; (4) distribution grid management; (5) Information and Communication Technology services; (6) use cases distinguishing goods transport and personal mobility, individual motorisation, and mobility as a service; (7) repair and maintenance; and (8) disposal, recycling, and retrofitting. The majority of these segments are touched upon in at least one regard by each selected case study.

National-level EV policy is basically corroborated and complemented by state- and city-level policy initiatives. Currently, 21 states have notified their EV policies (12) or published draft versions (nine) of the same, addressing different aspects of the EV value chain, institutional mechanisms, or the EV ecosystem more broadly.[2] In this chapter, state-level EV policies are presented wherever they are contextually relevant.

Section 11.2 specifies the methodological approach and defines the scope of analysis and databases, while introducing the foundations of a multidimensional framework of analysis that captures how business model innovations (BMI) cater to India's evolving EV ecosystem. The third section of this text presents the research findings of five selected BMIs. This section covers the business model of (1) Energy Efficiency Services Limited (EESL) – a government-owned company established to facilitate the implementation of different energy efficiency programmes across India, (2) Magenta Group (Maharashtra), (3) Telangana-based E-trio, (4) Sun Mobility – India's most prominent developer of battery swapping systems, and (5) Ola Electric, which is under the umbrella of mobility service provider, Ola Mobility. The last two companies are based in Bengaluru, Karnataka – the state that issued India's first subnational EV policy in 2017. Section 11.4 reflects on the role of businesses, governments, and the ecosystem more broadly in overcoming the barriers to EV adoption in India. It discusses the possibilities and limitations behind government ambitions to domestically capture the full EV value chain and presents policy recommendations derived from the analysis. Section 11.5 contains a summary and concluding thoughts.

11.2 Background and Method

In this chapter, we adopt Adner and Kapoor's (2010) narrow definition of an ecosystem's approach as a "network comprising a focal firm, its suppliers, its complementor firms, and customers." It is acknowledged that an EV ecosystem consists of multiple stakeholders beyond the firm, including finance, EV manufacturing, customers, after-sales services, charging infrastructure, government, and regulatory bodies (Soman et al., 2019). For the purpose

Table 11.2 Case studies: Business models in the EV ecosystem.

	Company	*Origin*	*Ecosystem*	*Business model strategy*	*Market presence*
1	EESL	Delhi	Public procurement and charging infrastructure	EV rental, sale, and charging for government officials	National/ Metropolitan India
2	Magenta	Maharashtra	EV supply equipment manufacturer and operator	Platform-based renewable electricity supply and charging service	National/ Metropolitan India
3	E-Trio	Telangana	EV supplier	Vehicle retrofitting	National/ Metropolitan India
4	Sun Mobility	Karnataka	Charging infrastructure service provider	Battery swapping	National/ Metropolitan India
5	Ola Electric	Karnataka	Mobility and charging service provider, e-2w supplier	(E-)mobility as a service (public and B2B), vehicle sales	Metropolitan India and international

of this research, however, BMIs are placed at the core of the analysis. Thus, instead of solely zooming in on the effects of public policy, EV adoption is approached through BMIs in an ecosystem context. The EV industry of India provides prime examples of an emerging business ecosystem in the early stages of its development. Accordingly, the case-study method (Yin, 2008) has been applied to better explain the role of BMIs in addressing the barriers to EV adoption. The ecosystem is captured as a sampling unit and business models as units of analysis, adopting Weiller and Neely's (2013) approach. As depicted in Table 11.2, five cases have been selected following an extensive literature review comprising industry reports, policy documents, company websites, and other news sources. Through a purposeful sampling method, information-rich, business-oriented cases in the Indian EV space have been selected that represent different and complementary parts of the EV value chain while addressing multiple aspects of the analysis framework.

Weiller and Neely's (2013) multidimensional framework for EV commercialisation distinguishes two broad categories – BMI around barriers to consumer adoption and enablers of EV ecosystem development. The framework further specifies each category through 11 aspects. In the Indian context, general barriers to EV adoption are identified in seven areas: (1) the lack of an EV mandate, (2) insufficient charging infrastructure, (3) too stringent or narrow conditions to avail subsidies, (4) high EV purchase costs and the dependence on battery imports, (5) an absence of adequate finance, (6) lack of public awareness,[3] and (7) inadequate availability of suitable EV models (GIZ, 2021; Kumar Tarei et al., 2021; Soman et al., 2019). For the purpose of this research, these findings are sufficiently close to Weiller and Neely's catalogue of business model evaluation criteria and will be discussed in more detail in Section 11.4 of this chapter.

However, after a critical review of these criteria, aspects had to be removed and added in order to reflect the actual challenges in India more pertinently. The criteria of "customer exposure to electricity price volatility" that Weiller and Neely bring up in the original framework does not resonate in the Indian context. Electricity tariffs in India are regulated by state electricity distribution companies. The Maharashtra State Electricity Distribution Company, for example, incentivises load shifting by offering a negative tariff (–1.5 INR/kWh) for overnight charging (10 pm–6 am) (Mahavitaran, 2020). Relevant business models such as that of Magenta, Sun Mobility, and Ola Electric also do not specify how exactly they handle electricity prices, that is, whether they fully pass on the price volatility to the consumer or charge a fixed price for their services. Electricity prices do not fall within the narrow scope of issues that can be addressed by private-sector BMIs and are thus removed from the framework.

In contrast, under the category of an enabler of the EV ecosystem development, the added aspect of "industrial policy support" allows us to assess the extent to which public policy is conducive to the respective business model. Furthermore, a third category was established dedicated to the environmental co-benefits of EV market formation, accounting for the trends towards a circular and sharing economy decoupled from extensively generating greenhouse gases (Buruzs & Torma, 2017; Greenblatt & Shaheen, 2015). Accordingly, the revised multidimensional framework of analysis captures the following three categories and related aspects:

A. BMI around barriers to consumer adoption

1. Reduces battery ownership costs
2. Reduces vehicle ownership costs
3. Absorbs (technical, market, financial, infrastructural) risk across the ecosystem
4. Facilitates long-distance travelling
5. Encourages change in consumer behaviour

B. Enablers of EV ecosystem development

6. Enables technological innovation
7. Formulates a clear business model strategy
8. Enables business model experimentation
9. Uses intelligent charging infrastructure
10. Pursues a service-oriented business model
11. Industrial policy support

C. Environmental co-benefits of EV market formation

12. Use case (passenger mobility and goods delivery as services)
13. Use of renewable energy to fuel the vehicle
14. ICE vehicle retrofitting

The five selected business cases have been assessed against 14 aspects on a 0–10 scale, with 0 representing no amplitude towards the aspect and 10 indicating full amplitude. The findings are inductively inferred from secondary data sources, including company websites, industry reports, and newspaper articles.

11.3 Results

This section presents the findings of BMIs for EV adoption in India on a case-by-case basis. Figure 11.2 illustrates overall results showing that a small number of Indian businesses is addressing – each to a certain degree – every single aspect of the framework. This suggests a conscious presence in the socio-technical innovation space that is shaping up quickly and promises to serve people, the planet, and profit simultaneously.

11.3.1 Energy Efficiency Services Limited

EESL is a company established under the Bureau of Energy Efficiency (BEE) of the Ministry of Power in 2009. It is the first state-owned company exclusively dedicated to the implementation of energy efficiency measures in India. The Ministry of Power mandated EESL to stimulate demand aggregation through an innovative business model of bulk procurement of EVs. By fully absorbing the entire upfront cost of the adoption of EVs and their infrastructure, EESL would deploy electric cars in government offices all over the country. In 2018, a number of Memoranda of Understanding (MoUs) were signed with state governments, including those of Delhi, Andhra Pradesh, Gujarat, Maharashtra, Telangana, and others. Under the National E-Mobility Programme of 2018, EESL completed the procurement process of 10,000 e-cars, of which more than 1,500 electric cars (15 per cent) are currently under deployment in government organisations across India. The vehicles are given on a lease or purchase basis to replace the existing petrol and diesel vehicles taken on lease by the government organisations. It is EESL's objective to replace the existing 500,000 government vehicles with EVs, while generating fuel savings and reducing carbon dioxide emissions. Figure 11.3 illustrates EESL business profile.

At the same time, EESL is developing EV charging infrastructure and has signed MoUs with multiple stakeholders across municipalities and DISCOMs for local assessment studies in order to set up 10,000 charging stations over the next two to three years, delivering on the FAME India scheme (ET, 2020).

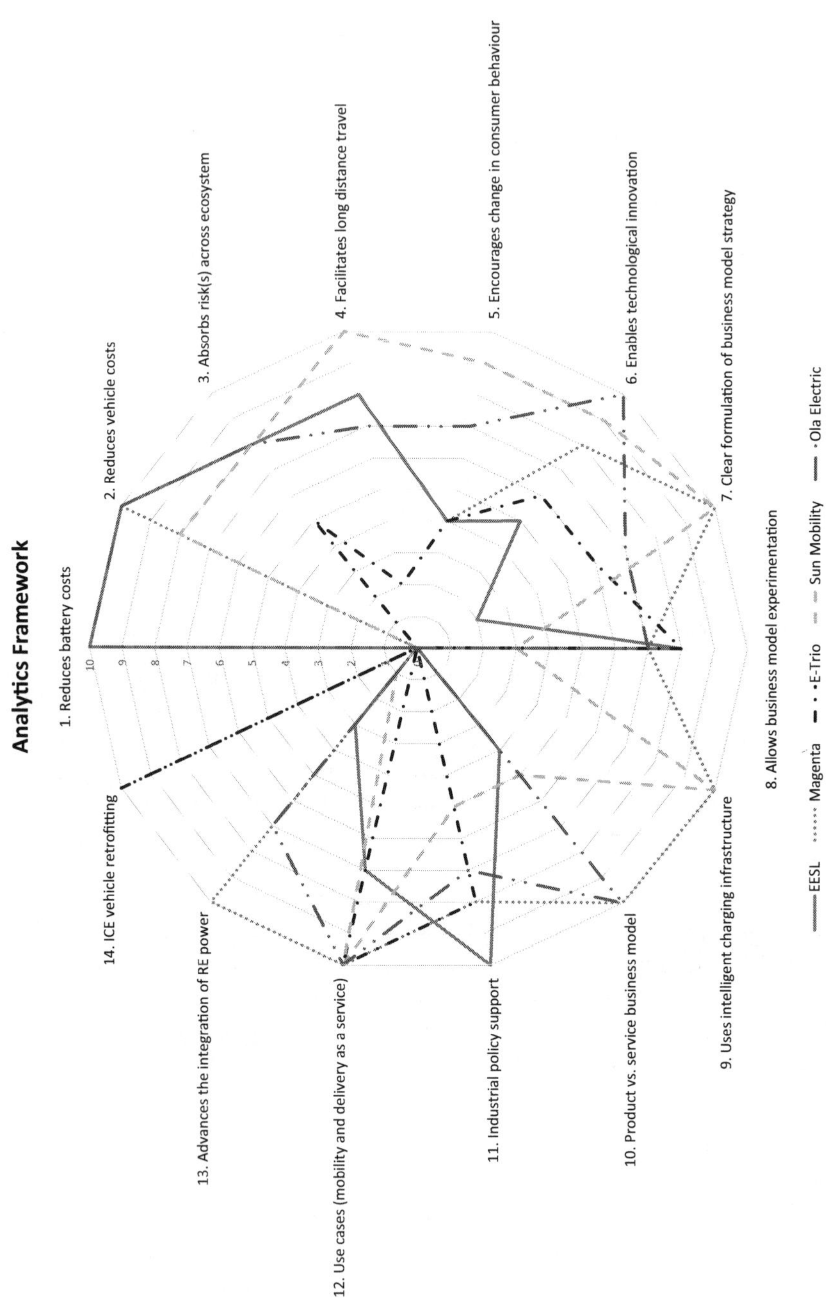

Figure 11.2 Framework analysis: An overview of all-case results.

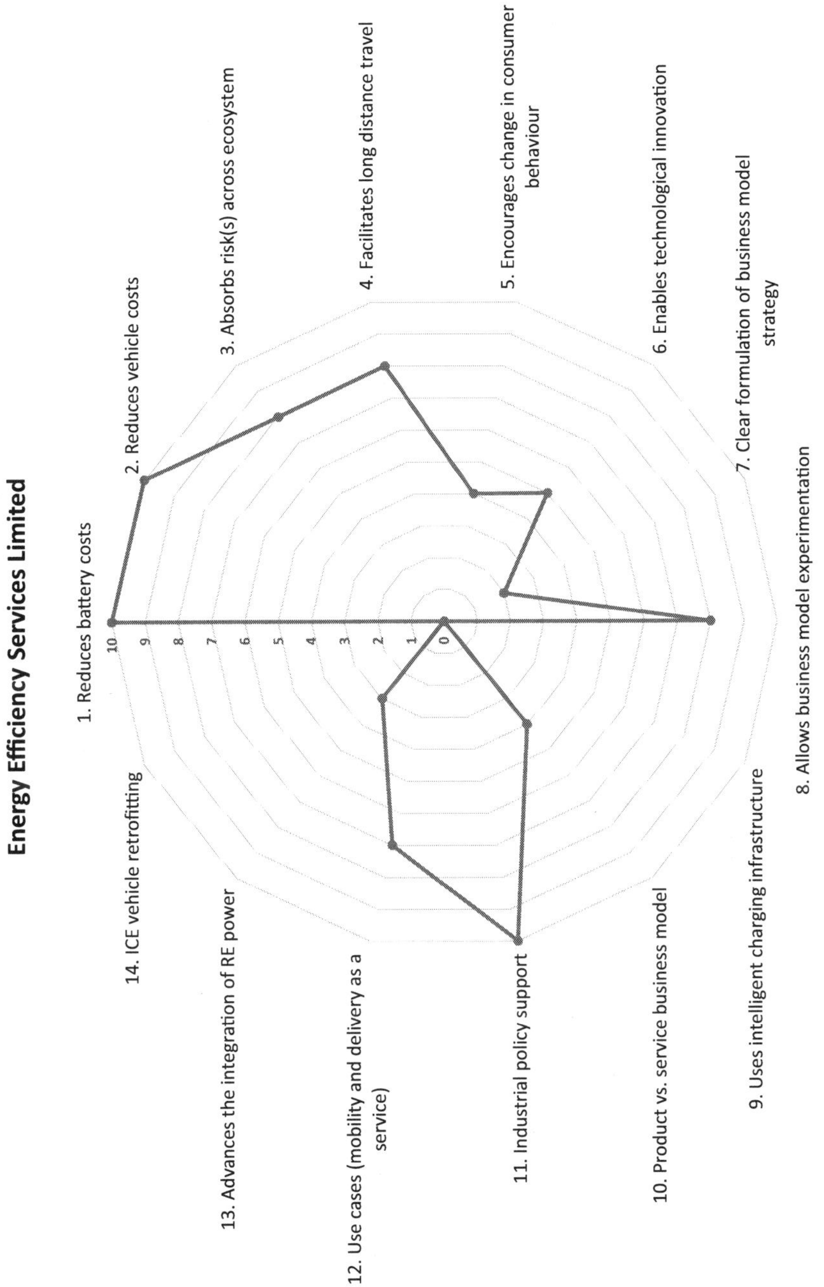

Figure 11.3 Framework analysis: The public procurement business model.

India's first public charging plaza at Chelmsford Club, New Delhi, for example, was inaugurated on July 20, 2020, and can charge 14 electric cars simultaneously. Regarding the ecosystem perspective, EESL is convinced that

> [t]his model will provide an impetus for Indian vehicle manufacturers, charging infrastructure companies, fleet operators, service providers, and the industry to gain efficiencies of scale and drive down costs, create local manufacturing facilities, and to grow technical competencies for the long-term growth of the EV industry in India. It will, in turn, enable the Indian EV manufacturers to emerge as a major global player.
> (EESL website)

11.3.2 Magenta Group (Maharashtra)

Established in 2017, Magenta Power has been revolutionising the EV space with a number of innovations starting with the launch of India's first EV portable charging station for offices and homes. Moreover, it has developed the ChargeGrid App for better orientation in the network of EV charging stations and for the facilitation of digital payment processing. The company has also established India's first solar photovoltaic (PV) charging station – a DC fast charger – in Turbhe, Navi Mumbai, Maharashtra (Tayal, 2018). Figure 11.4 illustrates Magenta's business profile.

Supported by state-owned Hindustan Petroleum Corporation Limited (HPCL), the Magenta Group consists of four verticals: (1) Magenta Power, (2) Magenta ChargeGrid, (3) Magenta Mobility, and (4) Magenta Informatics. Together they represent a corporate ecosystem that seeks to capture multiple EV business opportunities. "We see tremendous potential and growth opportunities to build a stronger EV ecosystem for India," says Kushal Banerjee, general manager – Business Development at HPCL (Magenta Group website).

Magenta Power is dedicated to a grid-connected clean energy supply. They install rooftop solar PV systems, solar PV carports, and wind turbines in combination with battery storage systems for homes, offices, and industries. In January 2020, Magenta Power opened an research and development (R&D) centre in Maharashtra, which would work on future technologies in the area of EVs, charging supply equipment, and smart data analytics. This new facility is set up to contribute to increased production rates and to accelerate the market readiness of AC and DC chargers for two-, three- and four-wheeled vehicles (Gupta, 2020a). Under the vertical ChargeGrid Flare, the company planned to set up 500 EV charging stations by the end of 2020, starting with HPCL retail outlets in major cities across India. Maxson Lewis, CEO of Magenta Power states,

> Understanding the existing EV challenges, what we are aiming to do is, instead of getting the car parked next to a charging station we bring the charging station to where the car is parked. This product is made specifically for India with low footprint and power situations. With

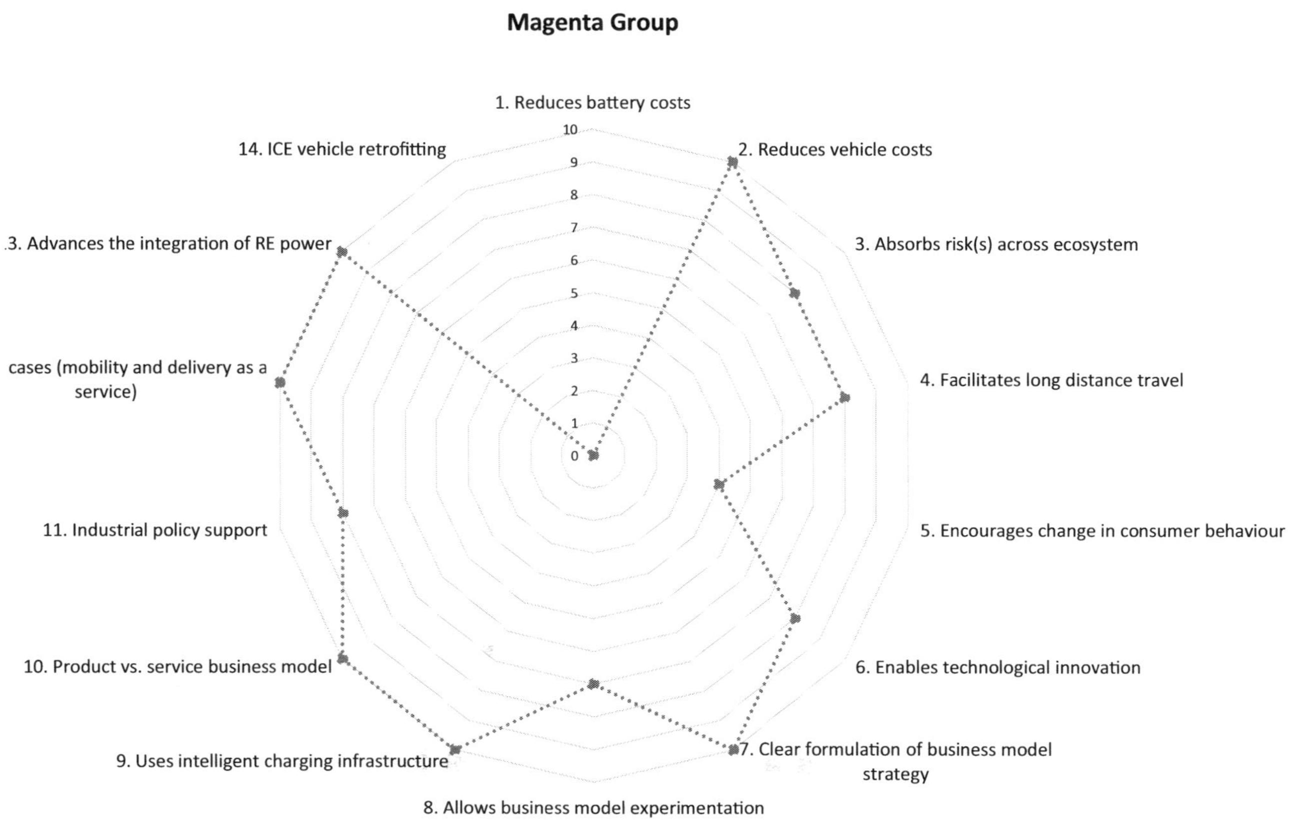

Figure 11.4 Framework analysis: The climate-neutral EV charging business model.

> ChargeGrid Flare, we can build a dense and convenient EV charging infrastructure at low cost, with less clutter to the existing city landscape.
> (Gupta, 2020b)

In March 2023, Magenta Mobility opened its largest charging depot in Bengaluru with 63 AC and 3 DC chargers. This marks the 35th charging depot in India with a plan for 14 more to be launched in FY 2023–2024 (FE, 2023). One example of Magenta's ingenuity is a new set of EV charging points that are incorporated in energy-efficient streetlamp columns in Mumbai and Delhi. Another example is ChargeGrid Prisma – a collaboration with several Residents Welfare Associations and Adani Electricity – providing shared EV charging solutions for residential societies in which not everyone has a dedicated parking lot (EV Reporter, 2021). With Magenta Mobility, the Group is also establishing a clean mobility platform for mobility as a service and goods transport. Magenta Informatics, on the other hand, serves as an IT solution provider for the generation, adoption, and growth of clean energy and mobility.

11.3.3 E-trio Automobiles (Telangana)

E-trio Automobiles is a Hyderabad-based vehicle retrofitting company established in early 2016. The team at E-trio has developed engineering design tools and manufacturing methods to convert conventional ICE vehicles into a wide range of customised lithium ion battery EVs from two- and three- to four-wheelers, including light commercial vehicles. The customer base consists of fleet owners providing employee mobility services and major logistics players like Amazon, Flipkart, BigBasket, IKEA, and Delhivery. E-trio is India's first company to have received a regulatory certification for retrofitting from the Automobiles Research Association of India (SiliconIndia, 2019).

With a team of more than 50 engineers and technicians and a facility of 50,000 square feet, it is capable of retrofitting 10,000 cars per year. The company describes itself as a catalyst for faster and more affordable EV mass adoption, all the while benefiting the environment more holistically. "Our vision is to create e-Mobility which inspires and nurtures pollution-free India" (E-trio website). Figure 11.5 illustrates E-trio's business profile.

According to E-trio, its major contribution to the EV industry in India is following the principles of a circular economy in which vehicles are kept in use as long as possible through multiple service lives. At the end of each life, materials in the car are recovered and regenerated. Following this practice, the on-road life of a vehicle can be increased beyond 15 years without adding another car and further congesting roads. "Retrofitting is a beautiful and scientific way to extend the life of your asset, it is a rejuvenation process," says Deepak MV, founder and CEO of E-trio. He adds,

> [I]f you look at the operating cost, an EV user would end up saving INR 3.5–4 per km (approximately 0.05 USD cents). So, the more you run,

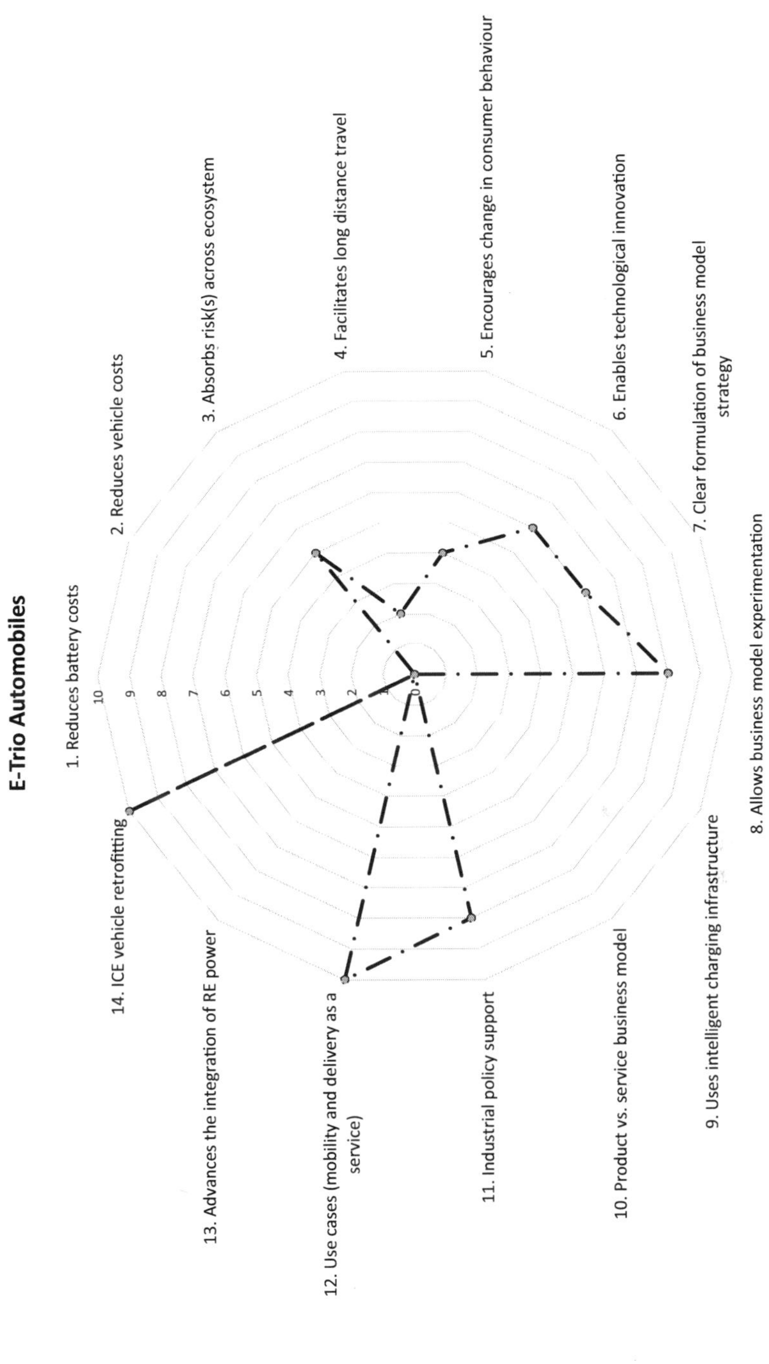

Figure 11.5 Framework analysis: The vehicle retrofitting business model.

> the more you save. We are looking at creating an extension of 5–7 years easily on the vehicle.
>
> (Sachdev, 2020)

At the same time, they "believe that affordable range is the key to helping electro mobility achieve a breakthrough soon," Sathya Yalamanchili, founder of E-trio.

At the policy level, E-trio demands regulatory support from the government. Currently, the national level FAME-2 policy has no provision or incentive for companies dedicated to vehicle retrofitting. "We know the kind of effort we've put into clearing the certification tests so that it meets a particular standard to operate. We understand the apprehension of the government, but they must support certified players like us. We want an equal playing field, where organised retrofitting is encouraged," concludes Deepak MV. The recently published Telangana State EV and Energy Storage Policy 2020–2030, however, provides an incentive of the state covering 15 per cent of the retrofitting costs (up to INR 15,000 per vehicle, approximately US$200) for the first 5,000 three-seater autorickshaws in Telangana (GoT, 2020).

11.3.4 Sun Mobility (Karnataka)

Sun Mobility is a Bengaluru-based company established in April 2017 by Chetan Maini – a pioneer in India's EV industry – and Uday Khemka as a 50:50 joint venture between the Maini Group and the SUN Group. Sun Mobility is India's most prominent player in the battery swapping space for electric three-wheelers and e-buses. The company provides a comprehensive set of solutions through an open-architecture energy infrastructure, including its own smart batteries that can be exchanged at its Quick Interchange Stations, mobile apps for drivers, and a web interface for fleet managers to interact with operators and locate vehicles. Sun Mobility is focusing on mobility as a service and has partnered with original equipment manufacturers (OEMs) and shared mobility providers such as Microsoft, Ashok Leyland, Uber, Piaggio, SmartE, Indian Oil Corporation, and Andhra Pradesh Economic Development Board. In August 2020, the scale of operations expanded to reach 11 Indian cities; the company moved over 300,000 passengers in electric three-wheelers and e-buses, covering more than two million kilometres without any tailpipe emissions. One year earlier, Sun Mobility and Ashok Leyland launched 18 Circuit-S buses with Ahmedabad Janmarg Limited, which operates the Bus Rapid Transit System of Ahmedabad. Battery swapping for the Circuit-S is supported by an automated battery swap station equipped with a robotic arm that can replace the 600-kilogram battery within three minutes (Wangchuk, 2019). Figure 11.6 illustrates Sun Mobility's business profile.

In 2020, the Ministry of Road Transport and Highways released a notification stating that electric two- and three-wheelers can be sold without a battery pack (EV Reporter, 2020). Bosch acquired a 26 per cent stake in Sun

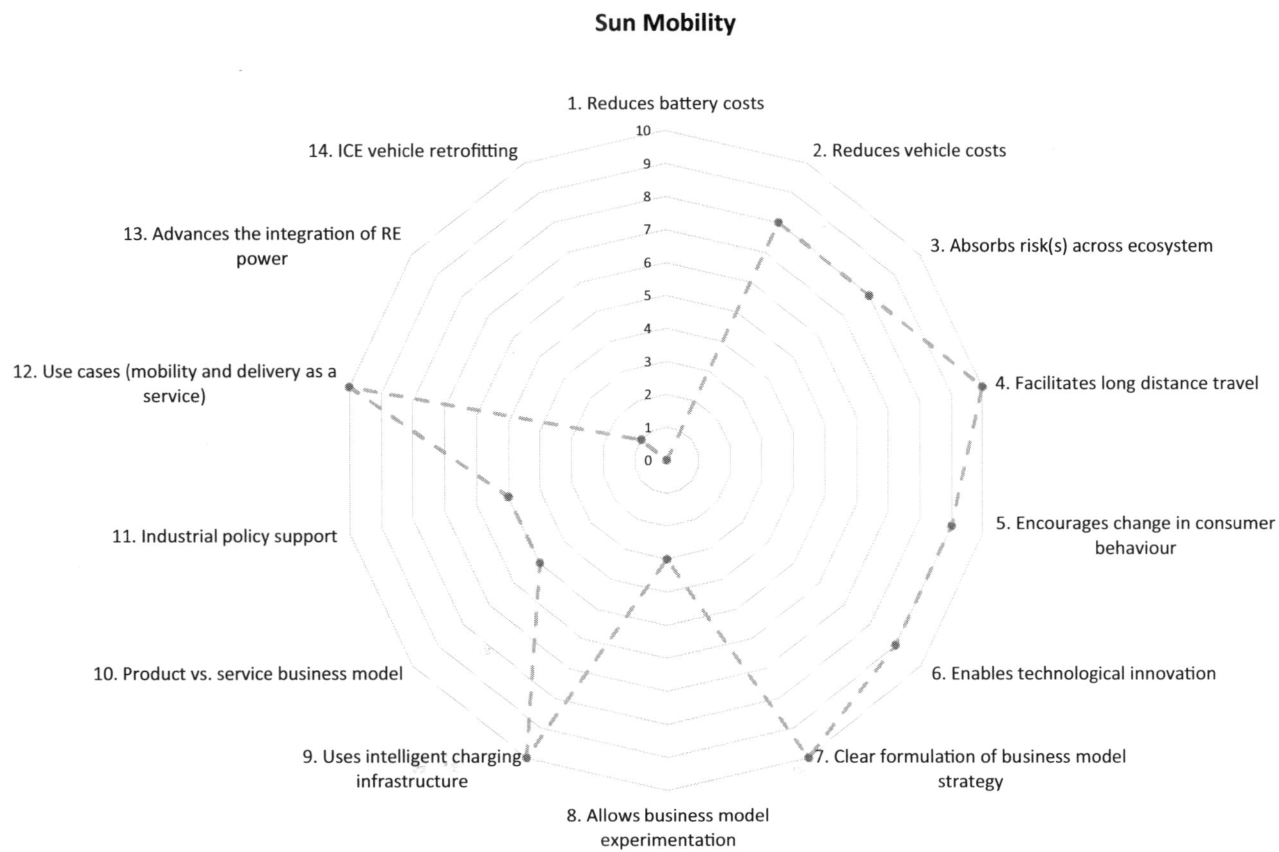

Figure 11.6 Framework analysis: The battery swapping business model.

Mobility (ET Auto, 2020), while the company launched battery swapping stations in Chandigarh and Bengaluru in collaboration with Indian Oil. In 2021, Sun Mobility announced its partnership with Tata Power-DDL, which provides electricity to more than 7 million people in Delhi. They jointly want to support the "Switch Delhi" campaign under the state government's innovative EV policy. It is Sun Mobility's vision to establish a network of interoperable energy infrastructure to accelerate the mass adoption of electric mobility. In 2021 alone, the company plans to set up 100 EV battery swapping stations across Bengaluru, aiming at powering one million EVs by 2025 across India (Sun Mobility website).

There are three major advantages of battery swapping with regard to overcoming the barriers to EV adoption that Sun Mobility capitalises on: (1) leasing vehicle batteries to the vehicle owner significantly reduces the upfront cost of the vehicle and helps avoid concerns over the resale value of the EV, (2) by decoupling the battery charging requirement from vehicle use, the driver saves a lot of time compared to wired EV charging, and (3) swapping stations recharge batteries in bulk in an air-conditioned environment, rendering cost-intensive fast charging redundant. Sun Mobility's digitally enabled "pay-as-you-go" system offers a highly convenient way to recharge an EV and makes them financially viable for the first time, thus fostering adoption (Sun Mobility website).

11.3.5 Ola Electric Mobility (Karnataka)

Ola Electric Mobility Private Limited is a Bengaluru-based company established in March 2017. It supplements the core business of India's largest mobility platform and one of the world's largest ride-hailing and food delivery companies. Ola was founded in December 2010 by Bhavish Aggarwal and Ankit Bhati as ANI Technologies Private Limited. The company would be dedicated to providing multi-modal mobility solutions for a billion-plus people in 250 plus cities in India and beyond.

Ola Electric Mobility specifically is working with various ecosystem partners, including vehicle and battery manufacturers, city authorities, power distribution companies, vehicle operators, and customers, to establish EV technology as a viable form of daily mobility. Building on the Nagpur pilot – India's first multi-modal EV project launched on 26 May 2017 – Ola invested INR 50 crore (approximately US$6.8 million at the time) in building charging infrastructure and integrating 200 electric cars and autorickshaws into its platform. In the process, it gathered valuable information about EVs, EV infrastructure, and ecosystem challenges (Arora & Raman, 2019). The project showed that a city could financially and environmentally benefit from EV adoption. "Installing solar rooftop at charging stations in Nagpur reduced average electricity expenses by 28%" (Arora & Raman, 2019). Figure 11.7 illustrates the company's business profile.

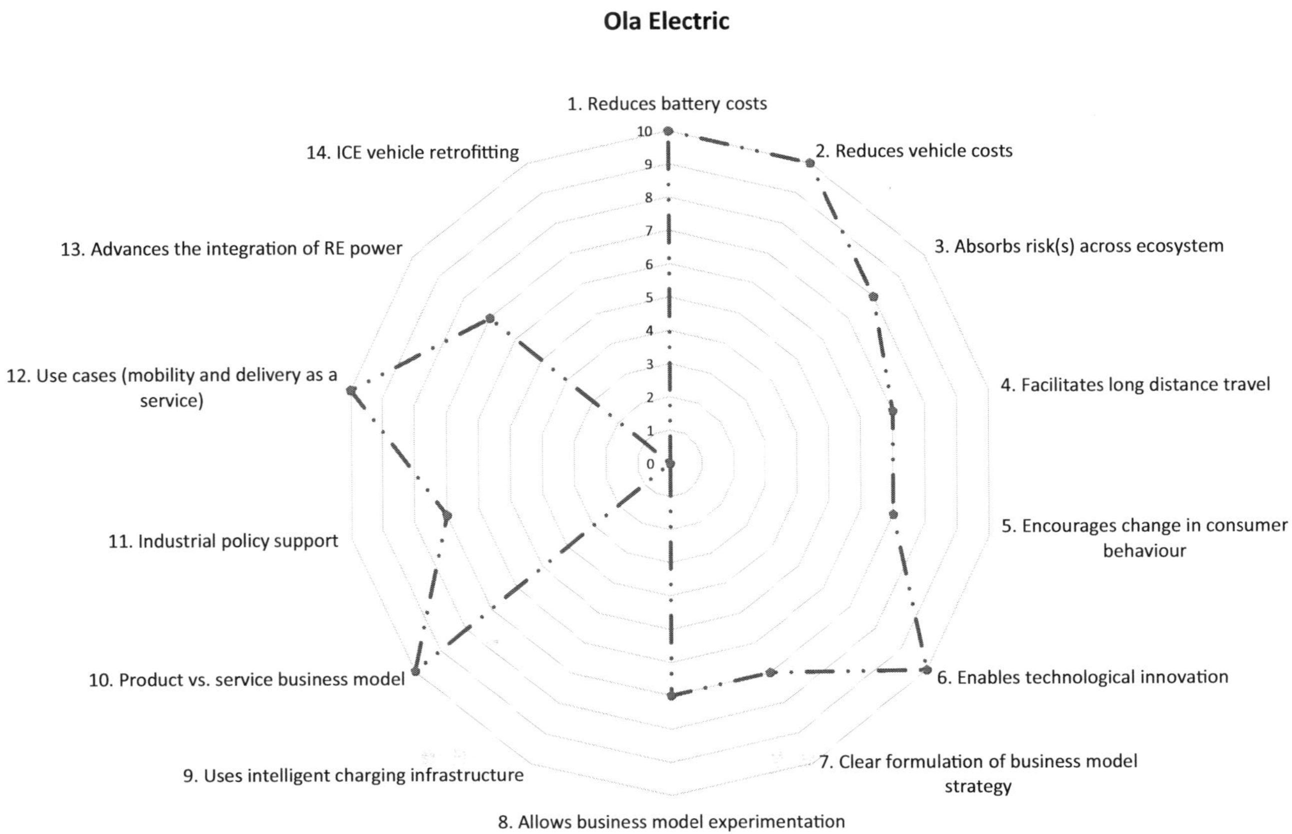

Figure 11.7 Framework analysis: From mobility service provider to EV manufacturer.

In May 2020, Ola Electric announced the acquisition of Etergo, an innovative, award-winning electric scooter OEM based in Amsterdam, the Netherlands. In December 2020, Ola signed an MoU of over INR 2,400 crore (US$327 million) investment with the Government of Tamil Nadu for setting up the world's largest e-scooter manufacturing plant in Hosur, Krishnagiri district. As of September 2020, roughly 17 per cent (14.7 GW) of India's entire renewable energy capacity was installed in Tamil Nadu, making the state a prime location for EV production and use (CEA, 2021). With an initial manufacturing capacity of over 2 million vehicles a year, the plant will create roughly 10,000 jobs. In September 2021, Bhavish Aggarwal revealed that the factory will become the world's largest production site entirely run by women.

> This is a significant milestone for Ola and a proud moment for our country as we rapidly progress towards realizing our vision of moving the world to sustainable mobility solutions across shared and owned mobility. This will be one of the most advanced manufacturing facilities in the world. This factory will showcase India's skill and talent to produce world-class products that will cater to global markets.

According to Ola, global annual two-wheeler sales are almost twice the volume of car sales. Through the strategic partnership with Etergo and the manufacturing site in Tamil Nadu, Ola Electric aims to lead and transform the 100-million-plus global two-wheeler market, including the 20-million-plus Indian two-wheeler stock, towards a clean, digital energy future (Ola Cabs' website).

11.3.6 Summary of Cases

All five BMIs presented go back no more than seven to eight years since their inception. Yet the findings suggest a strong and environmentally conscious presence in the global EV landscape, wherein each company plays an important role in the learning curve and ultimately in contributing to the establishment of a viable Indian EV ecosystem. It is noteworthy that a few of the cases already underwent significant expansions in their business model while others remain rather static. The maximum expansion conceivable is perhaps presented by Ola Mobility. Ola has not only established Ola Electric to provide electrified mobility services but also ventured into the domain of EV production in partnership with an international OEM. Similarly, the Magenta Group, with offerings in the mobility as a service space, envisions a market and value capture beyond its initial core business model of providing EV charging solutions powered by renewable energies. Sun Mobility has developed a battery swapping solution for e-buses and is likely to cater to other electric heavy-duty vehicles in the near future.

The extent to which cooperation or competition motivates these actors is not documented and interview requests remained largely unanswered. However, if these and other BMIs ideationally cross-pollinate and geographically expand, assisted by governments as described by Doblinger et al. (2019), chances are that India will become a leading global EV market by 2030.

11.4 Discussion and Policy Recommendations

When reflecting upon the barriers to EV adoption in India as presented by GIZ (2021) and others, it is striking that many of the listed issues seem to be missing their point. The lack of an EV mandate, for example, is not applicable. Power Minister Sri Piyush Goyal (2014–2017) and Minister of Road, Transport and Highways Sri Nitin Gadkari (since 2014) might not have been fully convincing with their 2017 plans, when they state that "[w]e are going to introduce electric vehicles in a very big way. . . . The idea is that by 2030, not a single petrol or diesel car should be sold in the country." However, it is encouraging to note that FAME phases I and II do corroborate the government's intentions to establish a solid ecosystem, ultimately leading to a significant increase in the numbers of EVs and supply equipment. FAME-I (2015–2019) allocated INR 343 crore for 278,000 EVs (plus 465 e-buses). Between January 2017, the start of available EV registration data, and April 2019, the end of FAME-I, more than 258,000 EVs have been registered countrywide. This, despite a lack of full spending transparency, suggests that a large part of the allocated funds effectively went into demand stimulation. This and the hockey stick shaped recovery in the sales statistics (Figure 11.1) create confidence in the achievability of objectives outlined in the successor programme even though the devastating second surge of the pandemic hit cities such as Delhi much harder than the first wave. The latest available countrywide data records another steep decline in EV sales in May 2021 even though it is not as low as the numbers one year earlier (Vahan Sewa Dashboard, 2023). Surprisingly, the EV sales recovery was much faster this time as depicted in Figure 11.1, resulting in a breakthrough moment with EV registration figures crossing the 5 per cent market share in 2021.

The claim that insufficient charging infrastructure would inhibit the uptake of EVs is also merely a half-truth. Currently, there are an estimated number of 6,600 installed and fully functional EV charging stations across India (MHI, 2023). The BEE provides a detailed list of charging stations per state or union territory. Under FAME phase II, more than 2,600 additional stations had been sanctioned. Given the predominance of electric two- and three-wheelers vehicles in the Indian market and the largely conventional household-level and small-business charging habits, these planned stations cater rather to mobility service providers. These include Lithium Urban Technologies and Ola Electric and the future market in the medium- to upper-sized EV segment. The above-presented outcomes show that efforts have been made towards demand stimulation and installation of charging stations. However, the large fraction of unspent funds under FAME-I seem to indicate that, with regard to technology platforms and pilot project, the conditions to avail subsidies are a tad too stringent or narrow. Yet to assess the validity of this claim is beyond the scope of this chapter.

The actual crux of the matter, where all analyses on barriers to EV adoption overlap, is that the purchase price of battery EVs is too high. Since the notification of the NEMMP in 2013, the government has deliberated upon strategies to lower the acquisition costs through reduced manufacturing costs. This could be achieved in three ways:

(1) Economies of scale associated with much higher volumes – the EESL approach.
(2) Lowering costs through technological developments – the R&D and Technology Platform approach.
(3) A higher degree of localisation – the government and Ola Electric approach.

Yet limitations to indigenise battery EV production remain, especially for large-scale, export-competitive battery and cell manufacturing. Mineral resources needed for battery cell manufacturing or electric motors such as Lithium, Graphite, and Neodymium are domestically scarce, which keeps the localisation potential at low levels (GIZ, 2021). At the same time, international demand is increasing steeply, creating a highly competitive market. Yet companies such as Li-Cycle have specialised in the recycling and recovery of critical battery resources from "electronic waste," which is an area to explore further from within India as well.

Domestic production of chassis and vehicle body, battery management systems, thermal management systems, and power electronics have very high localisation potential until 2030. These parts would still constitute 35 per cent of the vehicle cost. For any related R&D project, the government would contribute a maximum of 50 per cent of the project costs, leaving it to the business acumen of India's industrialists to accept or reject the EV challenge. The Indian automotive industry is a largely private-sector undertaking, hitherto considered influential when it comes to shaping norms and regulations and the rhythm and pace of their adoption (Balcet & Ruet, 2011). At the end of 2017, SIAM published a white paper through which it responded to the ministers' ambitions, stating that the introduction of EVs "in a big way" would be postponed to 2047 and the pressure to fully electrify vehicles was shifted to inner-city public transit (SIAM, 2017). Interestingly, FAME-II largely complies with the industry's dictum, announcing to put "greater emphasis on providing affordable and environmentally friendly public transportation options for the masses" (The Gazette of India, 2019).

The entry of international companies into the Indian vehicle market, with their own manufacturing know-how, might stir up Indian OEMs. More importantly, these companies may generate employment, and the government is well advised to assist these companies in succeeding in India because they, in turn, will help India to grow in more than just a self-reliant way.

11.5 Conclusion

This research shows that BMIs for EV adoption in India, despite being in its early stages, have started out with mature and environmentally conscious business objectives in mind. There has been hardly any major reported redesign or experimentation in the five presented BMI cases. The preference for electric sport utility vehicles over sedans might count as a minor yet costly adjustment. At the same time, all the firms explored in this chapter and their ecosystems at large have been going through a learning curve, whether that is reported or remains internal to the firm. The firmness of the Indian BMIs could be the result of cross-country learning effects preceding the commencement of the business, the local context, and the necessity to succeed at first try. While the profitability of each presented BMI remains a hidden measure, the (re-)configuration of the EV value chain towards higher levels of indigenisation promises increasing domestic value capture, which ideally translates into co-beneficial economies of scale for electric mobility.

Notes

1 India is a member of the Electric Vehicle Initiative of the International Energy Agency that encourages all its members to align with its EV30@30 campaign. This campaign pursues a target of electric vehicles comprising 30 per cent of the total vehicles sold in the country in 2030. The target is not mode-specific and remains clearly behind the Transformative Scenario outlined by NITI Aayog and RMI.

2 States/Union territories with EV policy (drafts) include Karnataka, Telangana, Maharashtra, Andhra Pradesh, Uttarakhand, Delhi NCR, Kerala, Bihar, Haryana, Uttar Pradesh, Tamil Nadu, Gujarat, Assam, Himachal Pradesh, Madhya Pradesh, Chandigarh, Punjab, Goa, Odisha, Meghalaya, and West Bengal.

3 In reverse, public awareness of safety and security issues also creates reservation towards EV adoption.

References

Adner, R. & Kapoor, R. (2010). Value creation in innovation ecosystems: How the structure of technological interdependence affects firm performance in new technology generations. *Strategic Management Journal*, 31(3): 306–333.

Arora, N. & Raman, A. (2019). *Beyond Nagpur: The Promise of Electric Mobility. Lessons from India's First Multimodal E-Mobility Project*. Ola Mobility Institute. Retrieved from: https://olawebcdn.com/ola-institute/nagpur-report.pdf (Last accessed: 17.01.2021).

Balcet, G. & Ruet, J. (2011). From Joint Ventures to National Champions or Global Players? Alliances and Technological Catching-up in Chinese and Indian Automotive Industries. Department of Economics and Statistics Cognetti de Martiis. Working Papers, 201109, University of Turin. Retrieved from: https://ideas.repec.org/p/uto/dipeco/201109.html (Last accessed: 20.01.2024)

Bieker, G. (2021). *A Global Comparison of the Life-Cycle Greenhouse Gas Emissions of Combustion Engine and Electric Passenger Cars. White Paper.* International Council on Clean Transportation (ICCT), Berlin. July 2021. Retrieved from: https://theicct.org/sites/default/files/publications/Global-LCA-passenger-cars-jul2021_0.pdf (Last accessed: 24.02.2021).

Bose Styczynski, A. & Hughes, L. (2019). Public policy strategies for next-generation vehicle technologies: An overview of leading markets. *Environmental Innovations and Societal Transitions*, 31: 262–272. https://doi.org/10.1016/j.eist.2018.09.002

Buruzs, A. & Torma, A. (2017). A review on the outlook of the circular economy in the automotive industry. *International Scholarly and Scientific Research & Innovation*, 11(6): 2017.

CEA (2021). *All India Installed Capacity (in MW) of Power Stations.* Central Electricity Authority as on 30.09.2020. Retrieved from: https://cea.nic.in/installed-capacity-report/?lang=en (Last accessed: 20.01.2024)

Doblinger, C., Surana, K. & Diaz Anadon, L. (2019). Governments as partners: The role of alliances in U.S. cleantech start-up innovation. *Research Policy*, 48(6): 2019. https://doi.org/10.1016/j.respol.2019.02.006

ET (2020). EESL to set up 10,000 EV charging stations in next three years. *The Economic Times Energyworld.* Retrieved from: https://energy.economictimes.indiatimes.com/news/power/eesl-to-set-up-10000-ev-charging-stations-in-next-three-years/74422674 (Last accessed: 02.12.2020)

ET Auto (2020). Bosch acquires 26% stake in Sun Mobility. *The Economic Time Auto.* 04.09.2020. Retrieved from: https://auto.economictimes.indiatimes.com/news/industry/bosch-acquires-26-stake-in-sun-mobility/77928929 (Last accessed: 08.05.2021)

EV Reporter (2020). *MoRTH Issues Notification on Sale of EVs without Battery Pack.* 13.08.2020. Retrieved from: https://evreporter.com/notification-on-sale-of-evs-without-battery-pack/ (Last accessed: 07.05.2021).

EV Reporter (2021). *Safe, Community Solution Needed for EV Charging in Residential Spaces – Maxson Lewis.* 03.04.2021. Retrieved from: https://bit.ly/3w98RmH (Last accessed: 08.05.2021)

FE (2023). Magenta mobility inaugurates the largest EV charging depot in Bengaluru. *The Financial Express.* 20.03.2023. Retrieved from: https://www.financialexpress.com/business/express-mobility-magenta-mobility-inaugurates-the-largest-ev-charging-depot-inbengaluru-3016515/ (Last accessed: 20.01.2024)

The Gazette of India (2019). *Scheme for Faster Adoption and Manufacturing of Electric Vehicles in India Phase II (FAME India Phase II).* Ministry of Heavy Industries and Public Enterprises, New Delhi. 11.03.2019.

Ge, M. & Friedrich, J. (2020). *4 Charts Explain Greenhouse Gas Emissions by Countries and Sectors.* World Resources Institute. February 2020. Retrieved from: www.wri.org/insights/4-charts-explain-greenhouse-gas-emissions-countries-and-sectors (Last accessed: 03.06.2021).

GIZ (2021). *Status Quo Analysis of Various Segments of Electric Mobility and Low Carbon Passenger Road Transport in India.* Deutsche Gesellschaft für Internationale Zusammenarbeit (GIZ) GmbH and Deloitte Touché Tohmatsu India LLP. Retrieved from: https://greenmobility-library.org (Last accessed: 04.02.2021)

GoI (2012). *National Electric Mobility Mission Plan (NEMMP) 2020.* Department of Heavy Industry, Ministry of Heavy Industries and Public Enterprises, Government of India. Retrieved from: http://dhi.nic.in/writereaddata/Content/NEMMP2020.pdf (Last accessed: 23.04.2020).

GoT (2020). *Telangana Electric Vehicle and Energy Storage Policy 2020–2030.* Government of Telangana. Retrieved from: https://tsredco.telangana.gov.in/Updates_2020/Telangana_EVES_policy_2020_30.pdf (Last accessed: 19.11.2020)

Greenblatt, J.B. & Shaheen, S. (2015). Automated vehicles, on-demand mobility, and environmental impacts. *Current Sustainable Renewable Energy Reports*, 2(2015): 74–81. https://doi.org/10.1007/s40518-015-0038-5

Gupta, U. (2020a). Magenta power opens R&D centre for EV chargers and energy informatics. *PV Magazine*. 14.01.2020. Retrieved from: www.pv-magazine-india.com/2020/01/14/magenta-power-opens-rd-centre-for-ev-chargers-and-energy-informatics/ (Last accessed: 05.11.2020)

Gupta, U. (2020b). Magenta power to roll out street lamp integrated EV charger. *PV Magazine*. 17.03.2020. Retrieved from: www.pv-magazine-india.com/2020/03/17/magenta-power-to-roll-out-street-lamp-integrated-ev-charger/ (Last accessed: 02.11.2020)

Kumar Tarei, P., Chand, P. & Gupta, H. (2021). Barriers to the adoption of electric vehicles: Evidence from India. *Journal of Cleaner Production*, 291(2021): 125847. https://doi.org/10.1016/j.jclepro.2021.125847

Mahavitaran (2020). *Revision of Electricity Tariff for FY20–21 to FY 2024–25*. Retrieved from: www.mahadiscom.in/consumer/wp-content/uploads/2020/04/Commercial_Circular_for_MYT_Order-3.pdf (Last accessed: 28.02.2021)

MHI (2023). *6,586 Operational Public EV Charging Stations in India*. Ministry of Heavy Industries. Government of India, New Delhi. 24.03.2023. Retrieved from: https://pib.gov.in/PressReleasePage.aspx?PRID=1910392 (Last accessed: 28.11.2023)

NITI Aayog and RMI (2017). *India Leaps Ahead: Transformative Mobility Solutions for All*. Retrieved from: www.rmi.org/insights/reports/transformative_mobility_solutions_india (Last accessed: 24.07.2020)

Sachdev, G. (2020). Why EV retrofitting is a step in the right direction. *Blue Circle*. Retrieved from: https://thebluecircle.co/2020/07/13/why-ev-retrofitting-is-a-step-in-the-right-direction (Last accessed: 13.07.2020)

SIAM (2017). *White Paper on Electric Vehicles: Adopting Pure Electric Vehicles – Key Policy Enablers*. Society of Indian Automobile Manufacturers. December 2017. Retrieved from: www.siam.in/uploads/filemanager/114SIAMWhitePaperonElectricVehicles.pdf (Last accessed: 03.07.2020).

SiliconIndia (2019). *E-Trio Automobiles: Certified EE Retrofitting Company. Special Issue on 'Start-up City'*. November 2019. Retrieved from: https://startup.siliconindia.com/vendor/etrio-automobiles-certified-ev-retrofitting-company – cid-10791.html (Last accessed: 05.05.2021)

Soman, A., Ganesan, K. & Kaur, H. (2019). *India's Electric Vehicle Transition: Impact on Auto Industry and Building the EV Ecosystem*. Council on Energy, Environment and Water (CEEW), New Delhi.

Tayal, M. (2018). *Magenta Power Sets-up India's 1st Solar Powered EV Charging Station*. Saur Energy International. 13.06.2018. Retrieved from: www.saurenergy.com/solar-energy-news/magenta-power-sets-up-indias-1st-solar-powered-ev-charging-station-mulls-to-expand-by-year-end (Last accessed: 01.11.2020)

Vahan Sewa Dashboard (2023). *Transport and Mobility Statistics of the Ministry of Road Transport and Highways, Government of India, New Delhi*. Retrieved from: https://vahan.parivahan.gov.in/vvahan4dashboard7 (Last accesed 15.12.2023)

Wangchuk, R.N. (2019). Adopted by Ahmedabad, 'battery-swapping' could revolutionize India's E-bus scene. *The Better India*. 09.02.2019. Retrieved from: www.thebetterindia.com/171736/india-ebus-electric-vehicle-costs-features-how-long-run/ (Last accessed: 07.05.2021)

Weiller, C. & Neely, A. (2013). *Business Model Design in an Ecosystem Context. Working Paper. Cambridge Service Alliance*. University of Cambridge, Cambridge, June 2013.

Yin, R.K. (2008). *Case Study Research: Design and Methods*, 4th edition. SAGE Publications, Inc, Thousand Oaks, CA.

12 Systems Thinking for Doughnut Cities

Ramya Natarajan, Kaveri Ashok, and Sarah Khan

12.1 Introduction

As of 2022, around 500 million people in India live in cities, which is 35 per cent of the country's population. By 2050, the urbanisation rate is expected to increase to over 50 per cent, according to estimates by the United Nations and the International Energy Agency (IEA, 2021). While the rate is still low compared to other countries, given India's large population and its developmental gaps, this translates to a massive increase in demand for urban infrastructure and energy in the coming decades. "Most of the buildings that will exist in India in 2040 have yet to be built" is one of the often-quoted statements made by the IEA (2021), highlighting the unique challenge ahead of India – economic growth paired with sustainable urbanisation.

Today, Indian cities are ranked quite poorly in the Global Liveability Index (EIU, 2022), which is calculated on the basis of safety, access to good healthcare, education, and other infrastructure. The urban "affordable housing" shortage in India is estimated to be around 19 million, according to the latest available data (Ministry of Housing and Urban Poverty Alleviation, Government of India, 2012), and is expected to double by 2050 if there is no increased effort towards meeting the gap (Kumar et al., 2021). Healthcare facilities, measures in terms of beds/1,000 people, is ~1.3 (0.8 from private hospitals and 0.5 from public), while the World Health Organization prescribes a minimum of three beds. Access to clean cooking fuel, water, transport, and power are other developmental challenges to overcome. Achieving these development objectives without making our cities less "liveable" necessitates strategies to stay within "planetary boundaries." But why can't India's urbanisation and growth be fossil-fuel-driven like cities in all developed countries so far? Why should India even consider an alternative growth strategy?

Without an attempt to stay within planetary boundaries, some of the reinforcing feedback loops pose the threat of irreversibly crossing critical tipping points in the multi-level earth system functioning. For example, if people with increased income continue to purchase and use multiple private cars, the issue of traffic congestion will get worse. If the cars have an internal combustion engine, this will contribute significantly to local air pollution and will

DOI: 10.4324/9781003281818-12

aggravate the urban heat island (UHI) effect leading in turn to more energy use. India has made ambitious pledges towards climate action culminating in the target for reaching net-zero emissions by the year 2070. All this necessitates a strategy that balances (urban) developmental goals and climate action in order to stay within "planetary boundaries."

If new buildings are not built with considerations of material input and aspects of climate resilience such as thermal comfort, this will make people more dependent on energy such as for ACs and more vulnerable to the impacts of climate change, which are already visible in the country. To break these vicious circles, the importance of systems thinking in urban planning and sustainable development is becoming increasingly important.

Synergies between climate action, sustainability, and improved quality of life can be found in using systems thinking in the form of causal loop diagrams (CLDs) and can be quantified using system dynamics modelling (Forrester, 1969). Systems thinking enables us to recognise the wider context of a problem, including the potential actions and interactions, the development of patterns, and the resulting consequences (Davidson & Venning, 2011; Kutty et al., 2020; Shrubsole et al., 2018). In other words, it improves our understanding of complex societal dynamics that arise in the urban environment. Using system thinking gives us insights into the underlying cause and effects, enabling us to better manage cities. Balancing the social, economic, and ecological needs of society suggests urban spaces to transform on the lines of a Doughnut economy (Raworth, 2017). The concept of a Doughnut City more specifically is driven by systematically capturing and understanding interdependencies between social goals, ecological impacts, economic aspirations, and global responsibilities as illustrated in Figure 12.1.

In this chapter, we combine aspects of systems thinking, modelling, and Doughnut Economics to conceptualise Doughnut Cities for India, where synergies between development and sustainability are identified and used in strategy formulation. In the following sections, we will first detail out the methodology used in developing a systems thinking framework and an overview of our model – the Sustainable Alternative Futures for India (SAFARI). We then present the results – insights from our CLD for urban systems, and decarbonisation scenarios using the SAFARI model. Finally, we discuss the best possible way forward for "Doughnut" urbanisation in India.

12.2 Methods

Models are used to visualise dynamic systems in order to better comprehend their behaviour and processes. These models replicate real-world scenarios, where the planning process and its players interact with the physical and institutional components of society, resulting in the creation of feedback or causal loops. A CLD is a representation of the system of interest where all causal relationships between key variables are mapped out graphically. Related variables are connected by an arrow; a positive sign indicates

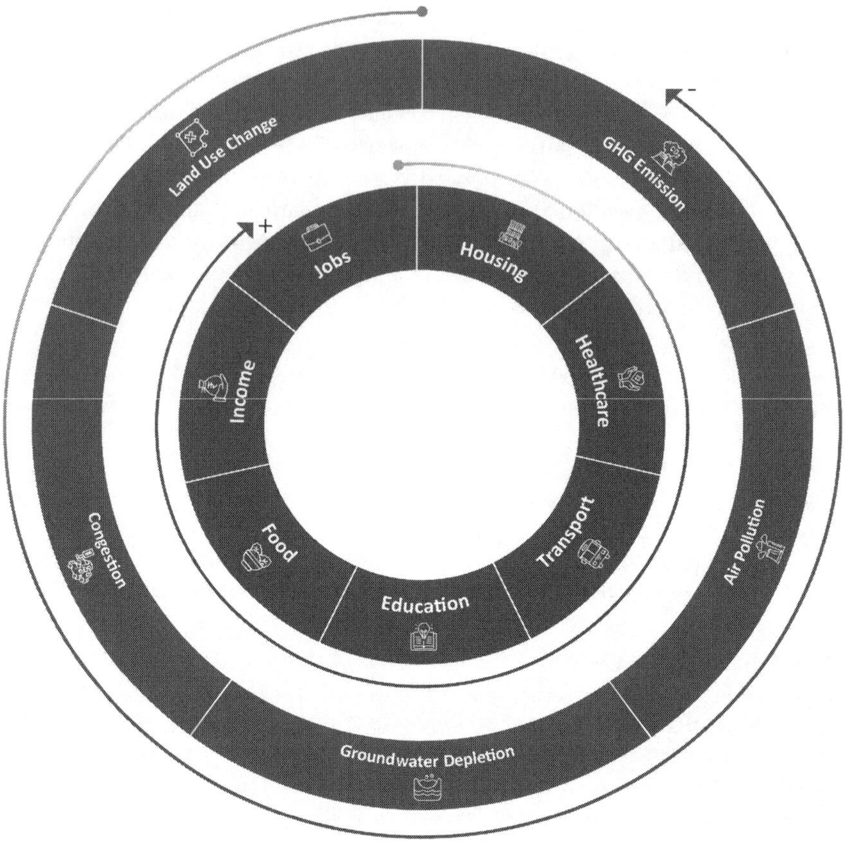

Figure 12.1 Conceptual diagram of a Doughnut City.

positive correlation, while a negative sign indicates inverse correlation. This is a tool used in systems thinking to understand the system and arrive at insights on leverage points and interventions to move the system towards desired change.

In this chapter, we focus on the energy demand and emissions aspects of cities and therefore depict only the relevant system variables. The SAFARI model (Ashok et al., 2021; CSTEP, 2021; Kumar et al., 2021) used in this analysis of energy and emissions is a system dynamics simulation model of India's development pathways running up to the year 2050. It is driven by demands arising from meeting development goals such as food and housing, in addition to macroeconomic parameters like gross domestic product and population. It follows a bottom-up approach and intends to provide an analytical framework that helps identify a sustainable long-term development strategy for India.

A Doughnut City needs to meet a comprehensive list of the targets, including the provision of food, housing, education, and healthcare to all its citizens without overshooting the ecological ceiling. The SAFARI model is developed based on this principle. The housing sector module of SAFARI focuses on the material and energy implications of meeting India's housing shortage. The primary objective of the module is to determine the shortage in urban housing of economically weaker sections, the lower income group, and shortage filling and reconstruction requirement of housing of the middle-income group (MIG) and higher-income group (HIG) (MoHUA, 2017b). The following are the factors considered for estimating housing shortage in the SAFARI model:

1. Dilapidation/Obsolescence, which includes both existing housing stock that is structurally too old or unfit for habitation, and any new housing stock that may be built to address the housing gap.
2. Congestion, which is the number of couples not having separate rooms to themselves. Congestion factor has been calculated using National Sample Survey data, which is estimated to be 18.42 per cent of urban households based on 2011 population census data (Ministry of Housing and Urban Poverty Alleviation,[1] Government of India, 2012).
3. Homelessness, which is estimated to be around 0.53 million as of latest available data (Ministry of Housing and Urban Poverty Alleviation, Government of India, 2012).
4. Percentage of housing stock reconstructed annually due to natural disasters.
5. Voluntary reconstruction of MIG/HIG houses.

If there is a land constraint for building the required stock, SAFARI allows for consideration of the densification or compact city scenario. Further, the land recovered from the demolition of dilapidated houses feeds back into available land. The total built-up area is calculated based on the existing housing stock, the total houses added from shortage filling, and the average built-up area for each housing category.

Through the SAFARI model, we explore the combined effect of urban densification, open green space, electrification of transport sector, electric cooking penetration, use of alternative materials for construction, and clean power in energy demand and emissions of Indian cities.

12.3 Results and Discussion

As countries develop, it has been observed that people move away from agriculture-based jobs typically associated with rural areas to the cities in search of better standards of living, resulting in an increase in urbanisation (van Ruijven et al., 2008). Increased urbanisation has a positive impact on overall economic growth. However, if the externalities of increased urbanisation are not well managed, this system would show various symptoms of overuse and might ultimately collapse. Figure 12.2 shows the complex interlinkages of the urban space with externalities such as air pollution, land-use change, traffic congestion, and GHG emissions contrasted with interventions

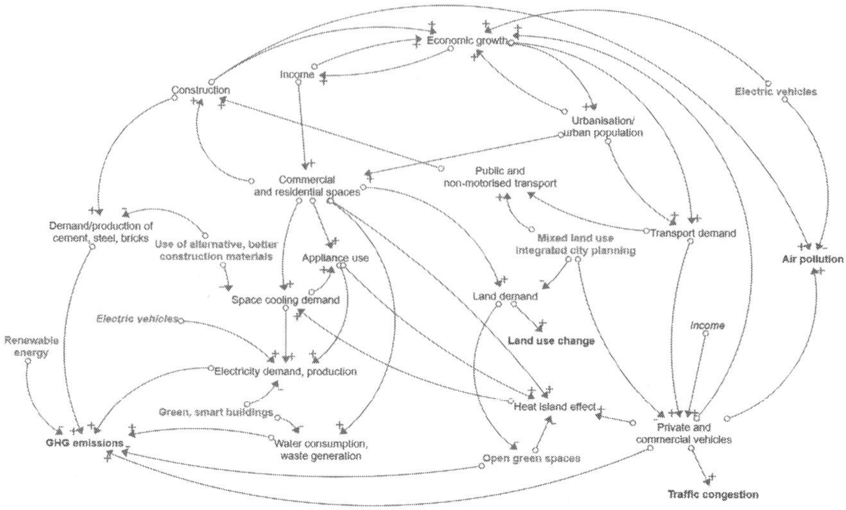

Figure 12.2 CLD of urban India with a focus on energy and emissions.

to help move towards sustainability such as renewable energies, electric vehicles, mixed land-use integrated in city planning, the use of alternative, better construction materials for msart buildings and other infrastructure, as well as open green spaces.

Growing population in urban areas would increase the demand for housing, healthcare and education facilities, other commercial spaces, and transport. This, in turn, increases the demand for construction, electricity, clean water, vehicles, and the underlying supportive fuels. These are petroleum fuels under a business-as-usual (BAU) scenario leading to an increase in externalities such as air pollution, greenhouse gas (GHG) emissions, traffic congestion, and groundwater extraction. If no attempt is made at decreasing these externalities, cities can easily become unliveable and even basic development goals and quality of life will be difficult to achieve.

To meet the growing demands of cities more sustainably, some interventions include – mixed land use integrated city planning (compact cities, increased public and non-motorised transport, increased open green spaces, etc.), using alternative better construction materials, increasing the shift to electric vehicles, increasing the adoption of renewable energy (RE) for power, and slowly replacing existing building stock with green and energy-efficient buildings. While it has been well established that these strategies would reduce negative externalities, it is not clear yet by exactly "how much." This is because some of these strategies come with trade-offs that also need to be accounted for. For example, a compact city will have taller buildings (to accommodate the population within lesser area), which need more steel overall for construction, and such a dense agglomeration could lead to increased UHIs which further increases space cooling demand in the building stock of a country with predominantly tropical climate. But a compact city will have reduced demand for motorised transport (and emissions caused by it) and a smaller footprint on land use. Thus, what are the net benefits of a compact city? How can we maximise these benefits and minimise the trade-offs? In the next section we quantify the impacts of each intervention in the CLD to identify the net environmental impact. We believe that this provides some direction for Indian cities to move towards sustainability and net-zero emissions.

12.3.1 Carbon Footprint of Doughnut Versus Regular Cities

The SAFARI model can be used to develop a comprehensive strategy for India's net-zero transition, taking into account its development goals, international commitments, and limited resources. Using the various levers of SAFARI discussed in the CLD, we have simulated scenarios to evaluate the environmental impact of a "Doughnut City" relative to a BAU scenario of a typical city.

In this scenario, we assume historical trends and practices to continue, resulting in urban sprawl. Urban sprawl has been linked to higher energy consumption, emissions, congestion, as well as a reduction in social distinctiveness and cohesiveness (Gavin, 2021). By 2050, the annual emissions from urban transport and buildings operation (cooking and appliances usage) are expected to exceed 560 Mt carbon dioxide equivalent (CO_2-e) under BAU conditions, which is more than double of today's emissions.

As a result, in order to improve the standard of living while respecting ecological boundaries, cities should aim to thrive and follow the Doughnut City development model rather than falling prey to a merely growth-oriented development paradigm. The implications of each lever and decarbonisation pathway for a Doughnut City are presented below.

12.3.1.1 Mixed Land Use and Integrated City Planning

Mixed land use, an idea that is widely accepted in urban development, has the potential to both enhance sustainability and provide desired quality of life to citizens. It typically entails increasing the extent to which different categories of land uses such as residential, commercial, and industrial are physically and functionally integrated in an attempt to make urban life more sustainable.

In SAFARI, we model this by increasing the floor space index (FSI or the height) of buildings as a proxy for densification/compact city, decreasing the average trip length based on city type, increasing the share of public and non-motorised transport, and increasing area under open green spaces. It has been observed that the compactness of cities increases cooling demand due to UHI effect (Kumar et al., 2021; Zhou et al., 2017). There is also an inverse correlation between UHI and green spaces in Indian cities (Imam & Banerjee, 2016). In addition to this, taller buildings necessitate a greater quantity of steel for their construction and an increased energy requirement for operating elevators and other systems, which can offset the emission savings from

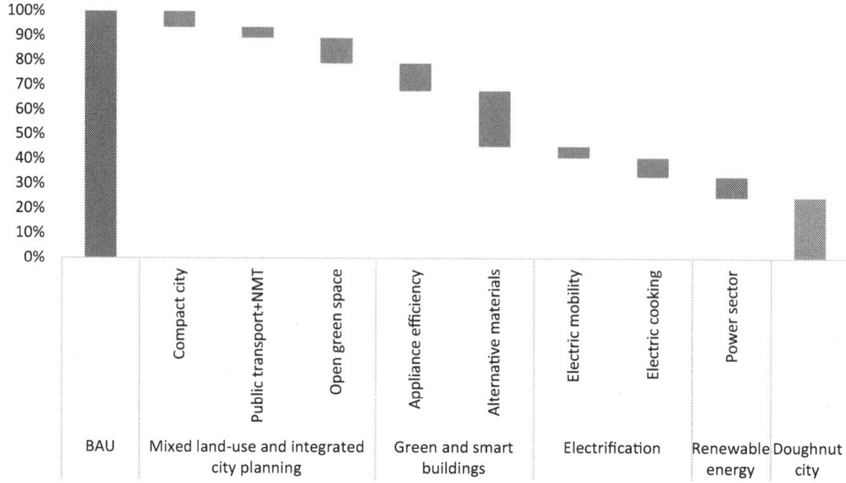

Figure 12.3 Annual urban GHG emissions reduction for a Doughnut City versus Regular BAU City.

densification and reduced transport demand (Ashok et al., 2021; CSTEP, 2021; Kumar et al., 2021). This requires compact cities to follow an integrated city planning approach, which according to SAFARI will result in a 21 per cent reduction in annual emissions as compared to BAU, as shown in Figure 12.3.

Increased open green spaces are one way of dealing with the UHI; a second approach is use of better construction materials which is considered in the next scenario described below.

12.3.1.2 Green and Smart Buildings

Green cities strive to create structures that are environmentally friendly, maximise energy efficiency, and utilise resources such as water and energy in a more optimal manner. Smart and efficient infrastructure follows a people-centric approach, prioritising the need of individuals. Energy-efficient infrastructure will significantly reduce emissions and provide a desired quality of life that aligns with the concept of a Doughnut City. Under this scenario, enhanced appliance efficiency, water recycling, and alternative materials for construction are considered. The most efficient alternative material for buildings under this scenario is autoclaved aerated concrete (AAC) blocks, which has low embodied emissions, energy, and thermal transmittance (Kumar et al., 2021). Other materials considered as alternatives to burnt clay bricks (BCBs) are hollow concrete, solid concrete, stabilised earth blocks, fly ash, and fly ash-lime-gypsum block.

Such a scenario could result in another 34 per cent reduction in annual emissions as shown in Figure 12.3. Use of alternative construction materials especially like AAC blocks has one of the biggest impacts on bringing down emissions – both embodied and operational – because of reduced demand for space cooling.

12.3.1.3 Transport Sector Electrification

Electrification of urban transport has the potential to drastically reduce air pollution, the UHI effect, and GHG emissions as well, if electricity is generated from REs. Data analysis has shown that Indian cities are dominated by two- and three-wheeler segments (TERI, 2019). Two-wheelers and three-wheelers are an attractive option for electrification due to their low cost and minimal infrastructure needs, allowing them to grow with minimal investment. Therefore, in this scenario, complete electrification of two-wheelers and three-wheelers by 2050, and moderate levels of electrification in other modes is assumed. Another assumption under the electrification scenario is electric cooking penetration in urban areas at around 75 per cent. Studies have demonstrated that the implementation of suitable policies can support the move from traditional cooking fuels to electricity (Kafle et al., 2019). Integrating strategies for clean cooking and electrification can lead to improved efficiency, as well as reduced electricity expenses. The electrification scenario can bring down annual GHG emissions by around 12 per cent if a BAU electricity grid is considered. The savings can be higher if increased RE is integrated, which we present in the next scenario.

12.3.1.4 Power Sector

For a successful transition to electric mobility and electric cooking, clean, reliable, and affordable electricity supply is essential. In this scenario, no new coal power plants, after 2025, are considered. The integration of RE sources into the powering of electric vehicles and e-cooking has the potential to reduce pollution and carbon emissions, as well as improve resource efficiency. In this scenario, the annual GHG emissions reduce by another 8 per cent.

When all these levers are combined, urban GHG emissions can reduce by almost 75 per cent annually, as shown in the waterfall chart (Figure 12.3). In addition to GHG emissions reduction, Doughnut Cities also have lower land and water footprints as explained in subsequent sections.

12.3.2 Co-benefits of a Doughnut City

12.3.2.1 Land Footprint

The urban built-up area has grown at a rate of 0.93 per cent between 2005–2006 and 2011–2012 and 2.3 per cent between 2011–2012 and 2015–2016 (MOSPI, 2017; EnviStats-India:Environmental Accounts, 2018), This trend is expected to continue with the increasing rate of urbanisation in the coming decades. Further, land increasingly becomes a precious resource because

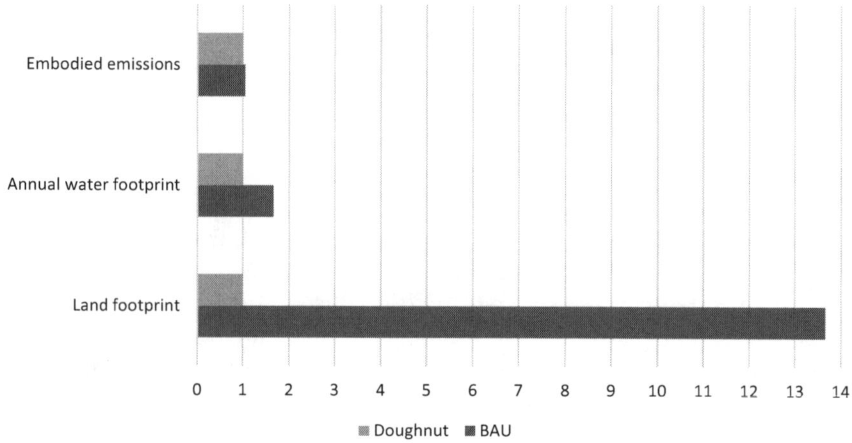

Figure 12.4 Land, water, and cumulative embodied emissions of BAU versus Doughnut Cities.

Note: Indicators are normalised for comparison.

of competition from other needs such as agriculture, forestry, and solar and wind installations. A Doughnut City can help reduce the land footprint of the urban built environment by almost 93 per cent, as shown in Figure 12.4. In particular, integrated land-use planning and compact urban development can have a positive effect on reducing the amount of pressure on land, both in the city and in nearby rural areas, so that agricultural and forest land are not consumed by urban sprawl.

12.3.2.2 Water Footprint

With depleting groundwater resources in many parts of India, water conservation practices become crucial for long-term sustainability. Buildings can be water-intensive. They are consuming significant amounts of water during their entire life cycle – "embodied water" in the construction materials used, water used on the construction site, and finally for operation once the buildings are occupied. Using alternative water-efficient construction materials and employing recycling practices for wastewater would be effective in reducing the water footprint in a Doughnut City, as illustrated in Figure 12.4.

12.3.2.3 Embodied Emissions

The construction rate of buildings in urban centres is expected to grow considerably and it has cascading impacts on the industrial sector. Cement, steel, and BCBs – which are currently the prevalent construction materials – are well known to be extremely energy- and emission-intensive. Embodied emissions of new construction are a good indicator that captures the emissions consequences of the building material choices. As discussed in the previous section, compact city planning with increased FSI implies taller buildings that require more reinforcing and therefore materials. The embodied energy (and associated emissions) of construction can be altered by selecting different combinations of building materials. The use of alternative material for construction, such as AAC blocks, with better thermal properties and lower energy requirements can significantly reduce space cooling demand and also lower the cumulative "embodied" emissions footprint for a Doughnut City. According to SAFARI results, cumulative "embodied" emissions can reduce by 4 per cent, which is around 600 Gt CO_2-e between now and 2050, equalling roughly 15 per cent reduction in annual emissions from construction.

12.4 Way Forward for Indian Cities

The making of a successful Doughnut City requires a number of ingredients to come together – the citizens (equivalent to the main ingredient of a doughnut – the flour), politicians and local administrators rising to the needs of the hour (like the yeast), researchers and thought leaders to provide evidence-based direction for policymaking (like the milk that binds everything together in a considerate way), while the local sugar, salt, nutmeg, or vanilla give the city its special taste. For the dough to become digestible, however, it needs finance (like the oil for its deep fry). For every Doughnut City in the making, regulations and policies are like the recipe that let the ingredients interact towards a desired outcome. In this section, we look at the ingredients and recipes of Indian cities through the existing institutional framework, the drivers of change, policies and gaps, and the way forward for urban India to develop as Doughnut Cities.

12.4.1 Institutional Framework

Planning, promoting, and coordinating environmental and climate change policies and programmes in India are the responsibility of the Ministry of Environment, Forests, and Climate Change (MoEFCC). In addition, the Ministry of Finance (MoF), the Ministry of External Affairs, and the NITI Aayog are involved in the development of climate policy. They collaborate closely with the MoEFCC to formulate an effective climate policy considerate of financial, developmental, and environmental agendas. To organise and oversee international climate finance, the MoF established the Climate Change Finance Unit. Under the direction of the MoEFCC, state-specific climate change action plans are handled by environment ministries and departments of science and technology in the union territories and at the state level. At the city level, the National Institute of Urban Affairs (NIUA) proposes two multi-stakeholder committees for effective climate action planning (NIUA, 2021).

1. Climate Coordination Cell at the local level with representatives from energy, waste, pollution, agriculture, water, public health, economics, and transport departments.
2. "Stakeholder's Committee" comprising department heads and administrators from Municipal Corporation, Town Planning Department, Regional Transportation Office, Electricity Board, Public Health Engineering Department, Builder's Association, Industrial Association, Indian Institute of Town Planners, local architects, educational and research institutes and non-governmental organisations.

Activating interactions with these groups could further enable the uptake of interventions discussed under Section 12.3.

12.4.2 Policy and Action Drivers

The system of global climate governance has undergone a significant change in recent years. The participation of several other organisations complements the established United Nations Framework Convention on Climate Change (UNFCCC) system. To address concerns related to climate change, for instance, networks, transnational organisations, and multilateral banks are increasingly collaborating with local and non-state players around the world. The Paris Agreement formally acknowledges the involvement of "non-party stakeholders" in the global effort to combat climate change and prepare for it. As a result, the role of sub-national and non-state actors inside the UNFCCC regime has gained more significance over time. Under the Non-state Actor Zone for Climate Action platform, the UNFCCC is formally documenting these initiatives. The latest Intergovernmental Panel on Climate Change report clearly points out ways how cities can precipitate systemic transformation – by striving for RE, environmentally friendly transportation, cleaner waterways, better waste management, and most crucially, investments in nature-based solutions as they continue to grow.

For the Indian context, Sastry and Michael (2019) identify five drivers or leavening agents of climate action at the city level covering aspects of multilevel government initiatives, responses to climate-induced extreme weather events, international partnerships, and civil society actions.

1. Central/state government mandated action: National acts, schemes, or programmes that are integrated into city developmental plans. For example, the Climate Smart Cities Assessment Framework (CSCAF), launched by the Ministry of Housing and Urban Affairs (MoHUA), aims to provide a clear road map for cities to combat climate change (MoHUA, 2021b).

2. Reactive action: Actions to mitigate climate-induced risks after experiencing extreme events. For example, Chennai's water resources department has several projects towards flood mitigation.
3. International partnerships: Action-driven partnerships with international organisations or funding agencies. For example, the Asian Cities Climate Change Resilience Network is working in Surat, Indore, and Gorakhpur to conduct stakeholder consultations and city risk assessments in order to develop climate resilience.
4. Political will: Initiatives steered through political willingness of the local leadership. For example, Brihanmumbai Municipal Corporation has devised the Mumbai Climate Action Plan to achieve net-zero carbon emission by 2050.
5. Citizen-led action: Civil society interventions are made that often lead to systemic changes in the local administration. For example, the conservation of lakes in Bangalore.

In our view, a better understanding of the multi-agent internal chemistry towards Doughnut City low-carbon development and its co-benefits, as described in Section 12.3, could enable and even speed up implementation through both political will (aligned with national government's mandate of net-zero) and deliberate reactive action.

12.4.3 Integrating Doughnut City Planning

In this section we carry out a comprehensive mapping of national-level policy and associated actors to build an understanding of the feasibility of the three central levers identified for transitioning to a Doughnut City.

12.4.3.1 Mixed Land Use and Integrated City Planning

The MoHUA has issued guidelines for state and local governing bodies through its National Transit-Oriented Development (TOD) Policy (MoHUA, 2017a), making suggestions for formulating state/city-level policies for promotion of TOD – efficient and high-density mixed land use. Further integrated planning of transport and land use in order to minimise the travel needs without compromising the needs of the population is envisaged in the National Urban Transport Policy (MoHUA, 2014a). There are also state/city-specific floor area ratio/FSI regulations which are used as a control tool to manage urban density dynamics (NITI Aayog, 2021). We further found national guidelines for Urban and Regional Development Plan Formulation and Implementation with respect to green and open space requirements in cities which are used by all urban local bodies (ULBs) (MoHUA, 2015). Evidently there is recognition at national scale regarding action levers for "mixed land-use and integrated city planning." However, the real challenge is for the state governments and the ULBs to comprehensively implement these guidelines in the local context. There is a significant financial crunch at the ULB level even for mega cities like Mumbai that makes implementation difficult that is for maintaining public green and open spaces, for example. The 15th Finance Commission recognised this fact and recommended grants on performance-based criteria for the ULBs (NITI Aayog, 2021). In order to fully integrate our research findings in moving towards mixed land use and integrated city planning, there needs to be more coordination between local and national governance, and of course sufficient local finance. Demonstrating the benefits of this transition in even one city or region could encourage action in others.

12.4.3.2 Green and Smart Buildings

The revised and amended National Building Code (NBC) provides guidelines on sustainability and energy-efficiency aspects including factors such as daylight utilisation, artificial lighting requirements, and heating, ventilation, and air conditioning design standards (BIS, n.d.). There is also the energy conservation building code developed the Bureau of Energy Efficiency (BEE) that provide minimum requirements for the energy-efficient design and construction of buildings. Currently, 22 states are at various stages of mandating the Energy Conservation Building Code (ECBC) in their building bye-laws. More recently, the BEE has also developed Eco-Niwas Samhita – a green building code for residential buildings. Furthermore, there are government incentives to enable the construction industry to facilitate the transition, which includes expedition of environmental clearances, additional floor-area-ratio, and reduction in property tax. The promotion of green buildings is also a key indicator in the CSCAF under the Smart City Mission. As of 2022, India has achieved 9.75 billion square feet of green building footprint spread across the entire country, which is the second highest green building footprint in the world (The Hindu, 2022). This suggests that policy supporting green buildings is enabling the transition fairly successfully. However, considering that the majority of urban infrastructure is yet to be built, we have a long road ahead. Building bye-laws are state subjects and therefore implementation in city-level is dependent on state regulations. At the local level, key actors are the town planning authorities, which ultimately ensure compliance and the Energy Auditors Board, as mandated by the ECBC which provides consultation and carries out the building approval process (NIUA, 2022). For appliance efficiency, the most significant government measure is the Standards and Labelling programme anchored in the BEE. It has been made mandatory for most of the common household appliances and is periodically updated to ensure the promotion of the most efficient technology (BEE, 2020). With respect to water recycling in cities, the Atal Mission for Rejuvenation and Urban Transformation (AMRUT 2.0) intends to promote the circular economy of water by creating city-level water balance plans for the 500 AMRUT cities, emphasising the recycling and reusing of treated sewage, revitalising water bodies, and water conservation (MoHUA, 2021a). In addition to policy support and regulatory measures, complementary approaches such as behavioural nudges can lead to more effective implementation of energy and water conservation objectives.

12.4.3.3 Electrification and RE

The National Electric Mobility Mission Plan is a comprehensive road map that targets the entire value chain of promoting electric vehicles. The Faster Adoption and Manufacturing of Electric and Hybrid Vehicle Scheme is the central government's two-phase incentive scheme promoting primarily small electric vehicles and public transport and has demonstrated success with record EV sales since the year 2022. According to Mohile (2022), EV sales are likely to exponentially increase fourfold by 2025. Besides, many states have specific EV targets and incentives in the form of road tax exemption and additional subsidies. As per the MoHUA guidelines, City Comprehensive Mobility Plans are expected to have a section dedicated to low carbon mobility plans with detailed emissions and pollution estimates from the transport sector (MoHUA, 2014b).

Similarly, cooking is a significant source of fossil fuel consumption in cities and the electrification of cooking will ultimately depend on the availability of reliable and cost-effective electricity. The power sector has ambitious targets of renewable integration and is well on track to achieve the enhanced targets under the renewed NDCs. At the city level, the Ministry of New and Renewable Energy is implementing a number of schemes, including setting up of solar parks, rooftop solar, solar street lights, and the like, towards establishing at least one "solar city" per state (MNRE, 2021).

Evidently, the Indian policy landscape is already conducive to almost all the levers identified in the Doughnut City model. However, there is definitely scope of improvement – for example by including complementary actions that emphasise behavioural interventions and mechanisms for consolidating funds at the ULB level.

12.4.4 International Examples

Cities are widely recognised as fertile grounds for sustainable transformation worldwide. There are successful examples for specific interventions towards achieving sustainability that offer lessons on complementary action that facilitate implementation of these interventions or levers.

The city of Portland in the United States has been implementing a Home Energy Score programme since 2018. This policy is designed to stimulate consumer demand for energy-efficiency upgrades and reduce carbon emissions by mandating the new homes to obtain and disclose their Home Energy Scores – which benchmarks household energy performance compared to others in similar climate zone. The policy enabled transparency in terms of the possible costs for the buyers and thereby incentivised sellers to make energy-efficiency improvements. The market has by now achieved almost full compliance and the economic and environmental benefits accrued have encouraged other cities like Berkeley and Austin to follow suit (Home Energy Scores Across the Country – Portland Home Energy Score, 2021).

The city of Copenhagen provides successful example in the context of sustainable urban transportation. Nearly half of Copenhagen's residents commute by bicycle, outnumbering cars by a factor of over five. A culture of bicycling has been fostered by constructing cycle superhighways with plenty of amenities along the way, such as traffic lights, air pumps, safer junctions, and city management (Fleming, 2018).

Amsterdam's local government and textile supply chain managers have taken the initiative to make their fashion industry more sustainable, signing a denim deal to produce 3 billion garments with 20 per cent recycled materials by 2023. Furthermore, a common repair shop has been set up for brands to encourage people to get their old jeans repaired rather than throwing them away, thus reducing the industry's ecological impact (Warren, 2021). This initiative works within both the local/ecological and global/ecological aspects of the Doughnut Model, making it a prime example of sustainable fashion.

Finally, the city council of Wellington in New Zealand has a wide variety of initiatives involving the local business communities and residents in order to reshape the city for a better future in response to climate change and the increasing need for housing. The council, with its continuous engagement with citizens through many forums, plays a central role in fostering innovation and driving climate action in the communities (WCC, 2021).

These success stories from other countries could inspire us to design and implement the interventions defining Doughnut Cities.

12.5 Conclusion

As per the 2011 census, 31 per cent of India's population was urban. This is projected to surpass 40 per cent by 2030, according to the NIUA (2016). Although it can bring a variety of economic advantages, speedy urbanisation also presents significant challenges. As India transitions from a rural to an urbanised society, energy consumption and associated GHG emissions are expected to rise. In 2021, India's per capita GHG emissions were 1.9 metric tonnes, with an annual 6 per cent increase over the past decade (Bhattacharya, 2022). This process presents both a challenge and an opportunity to promote sustainable development through the judicious use of technology and careful planning. We must rethink, reimagine, and revamp our approach to planning cities and towns in India in order to ensure their economic, environmental, and social harmony. In this chapter, we have presented the idea of a Doughnut City as a conceptual framework for the sustainable transformation of Indian cities. Using our simulation model, SAFARI, we also presented the implications of sustainability interventions that can help transition towards a Doughnut City such as mixed land use and integrated city planning, green and smart building infrastructure, electrification, and the use of RE.

The way a city is structured has a major impact on its transition towards a sustainable and low-carbon economy. Urban planning, including more efficient use of space and materials for building, combined with changes in transportation modes, is essential for reducing emissions in urban areas. We believe that the feasibility of implementation ultimately comes down to the political will of the local leadership and the capacity of ULBs to chart the way forward. The Doughnut City conceptual framework provides a systemic lens through which all the individual components, interventions, and policy actions can be brought together in a meaningful way to achieve long-term sustainability in our urban spaces.

Note

1 The Ministry of Housing and Urban Poverty Alleviation was created in 2004 and merged with the Ministry of Housing and Urban Affairs in 2017.

References

Ashok, K., Natarajan, R., Kumar, P., Sharma, K. & Mathur, M. (2021). Sustainable alternative futures for agriculture in India – the energy, emissions, and resource implications. *Environmental Research Letters*. https://doi.org/10.1088/1748-9326/abf0cd

BEE (2020). *Standards & Labelling*. Bureau of Energy Efficiency. Retrieved from: https://beeindia.gov.in/content/standards-labeling (Last accessed: 08.01.2023)

Bhattacharya, S. (2022). *Report At COP27: India Records Highest Emission Increase Among Top Global Contributors*. www.outlookindia.com/international/report-at-cop27-india-records-highest-emission-increase-among-top-global-contributors-news-236452

BIS (n.d.). *National Building Code*. Bureau of Indian Standards. Retrieved from: www.bis.gov.in/standards/technical-department/national-building-code/ (Last accessed: 08.01.2023)

CSTEP (2021). *Energy and Emissions Implications for a Desired Quality of Life in India via SAFARI*. Retrieved from: https://cstep.in/drupal/sites/default/files/2021-07/CSTEP_SAFARI_Final_Report_July2021.pdf (Last accessed: 03.08.2021)

Davidson, K.M. & Venning, J. (2011). Sustainability decision-making frameworks and the application of systems thinking: An urban context. *Taylor & Francis Online*. https://doi.org/10.1080/13549839.2011.565464

EIU (2022). *The Global Liveability Index 2022*. Economist Intelligence Unit. Retrieved from: www.eiu.com/n/campaigns/global-liveability-index-2022/ (Last accessed: 03.01.2023)

EnviStats-India 2018: Environmental Accounts (2018). *Land-The Base, The Foundation*. Retrieved from: https://mospi.gov.in/documents/213904/0/3_Chapter+1-+Land+%282%29.pdf/21d49430-e4df-a205-61e6-62143f3075ee?t=1602851827983 (Last accessed: 20.12.2022)

Fleming, S. (2018). *What Makes Copenhagen the World's Most Bike-Friendly City?* World Economic Forum. Retrieved from: www.weforum.org/agenda/2018/10/what-makes-copenhagen-the-worlds-most-bike-friendly-city/ (Last accessed: 20.01.2023)

Forrester, J. (1969). *Urban Dynamics*. MIT Press, Cambridge, MA.

Gavin (2021). *Doughnut Cities*. Retrieved from: https://smartcityhub.com/governance-economy/doughnut-cities/ (Last accessed: 03.12.2022)
The Hindu. (2022). Green building footprint. *The Hindu*. Retrieved from: www.thehindu.com/life-and-style/homes-and-gardens/green-building-footprint/article66094823.ece (Last accessed: 04.11.2022)
Home Energy Scores Across the Country – Portland Home Energy Score (2021). Retrieved from: https://orderhomeenergyscore.com/2021/04/22/home-energy-scores-across-the-country/ (Last accessed: 20.01.2023)
IEA (2021). *India Energy Outlook 2021*. International Energy Agency, Paris. Retrieved from: https://iea.blob.core.windows.net/assets/1de6d91e-e23f-4e02-b1fb-51fdd6283b22/India_Energy_Outlook_2021.pdf (Last accessed: 03.12.2022)
Imam, A.U.K. & Banerjee, U.K. (2016). Urbanisation and greening of Indian cities: Problems, practices, and policies. *Ambio*, 45(4): 442–457. https://doi.org/10.1007/s13280-015-0763-4
Kafle, S., Mishra, A.K., Maharjan, A. & Panday, A.K. (2019). Moving from fossil fuels to electric cooking: An option for clean cooking access in Nepal. *Journal of the Institute of Engineering*, 15(3): 3. https://doi.org/10.3126/jie.v15i3.32174
Kumar, P., Natarajan, R. & Ashok, K. (2021). Sustainable alternative futures for urban India: The resource, energy, and emissions implications of urban form scenarios. *Environmental Research: Infrastructure and Sustainability*. https://doi.org/10.1088/2634-4505/ac048e
Kutty, A.A., Abdella, G.M., Kucukvar, M., Onat, N.C. & Bulu, M. (2020). A system thinking approach for harmonizing smart and sustainable city initiatives with United Nations sustainable development goals. *Wiley Online Library*. https://doi.org/10.1002/sd.2088
Ministry of Housing and Urban Poverty Alleviation (2012). *Report of the Technical Group on Urban Housing Shortage 2012–17*. Government of India, New Delhi.
MNRE (2021). *One Solar City in Each State/UT*. Ministry of New and Renewable Energy, Government of India. Retrieved from: https://solarrooftop.gov.in/notification/128_notification.pdf (Last accessed: 20.12.2022)
Mohile, S.S. (2022). EV sales likely to jump by over 4X by 2025 in India: Deloitte Touche. *Business Standard India*. Retrieved from: www.business-standard.com/article/automobile/ev-sales-likely-to-jump-by-over-4x-by-2025-in-india-deloitte-touche-122090901029_1.html (Last accessed: 03.01.2023)
MoHUA (2014a). *National Urban Transport Policy*. Ministry of Housing and Urban Affairs, Government of India, New Delhi.
MoHUA (2014b). *Toolkit for Comprehensive Mobility Plan – Revised*. Ministry of Housing and Urban Affairs, Government of India, New Delhi.
MoHUA (2015). *Urban and Regional Development Plans Formulation and Implementation (URDPFI) Guidelines*. Retrieved from: https://mohua.gov.in/upload/uploadfiles/files/URDPFI%20Guidelines%20Vol%20I(2).pdf (Last accessed: 20.12.2022)
MoHUA (2017a). *National Transit Oriented Development (TOD) Policy*. Ministry of Housing and Urban Affairs, Government of India, New Delhi.
MoHUA (2017b). *Pradhan Mantri Awas Yojana-Urban (PMAY-HFA(Urban)*. Ministry of Housing and Urban Affairs, Government of India, New Delhi. Retrieved from: https://pmaymis.gov.in/ (Last accessed: 29.01.2023)
MoHUA (2021a). *Atal Mission for Rejuvenation and Urban Transformation 2.0*. Ministry of Housing and Urban Affairs, Government of India, New Delhi. Retrieved from: https://mohua.gov.in/upload/uploadfiles/files/AMRUT-Operational-Guidelines.pdf (Last accessed: 03.12.2022)
MoHUA (2021b). *ClimateSmart Cities*. Ministry of Housing and Urban Affairs, Government of India, New Delhi. Retrieved from: https://smartcities.gov.in/climatesmart_cities (Last accessed: 03.12.2022)

MOSPI (2017). *The SEEA Ecosystem Accounts for India. POLICY BRIEF.* Ministry of Statistics & Program Implementation, Government of India. Retrieved from: https://mospi.gov.in/documents/213904/1319361/NCAVES+-+INDIA+-+Policy+Brief+-+FINAL+WEB.pdf/0584191a-9680-fa4f-561c-1a1f976eb96f?t=1631781688471 (Last accessed: 04.12.2022)

NITI Aayog (2021). *Reforms in Urban Planning Capacity in India.* Retrieved from: www.niti.gov.in/sites/default/files/2021-09/UrbanPlanningCapacity-in-India-16092021.pdf (Last accessed: 04.01.2023)

NIUA (2016). *Urbanisation and Urban Sprawl: A Perspective on India's Urban Growth.* Retrieved from: https://niua.org/tod/todfisc/book.php?book=1&section=2 (Last accessed: 23.01.2023)

NIUA (2021). *City Climate Action Plan – Training Manual.* National Institute of Urban Affairs. Retrieved from: https://niua.in/c-cube/sites/all/themes/zap/pdf/climate-action-plan.pdf (Last accessed: 04.01.2023)

NIUA (2022). *Promotion of Green Buildings – Training Manual.* Retrieved from: https://niua.in/c-cube/sites/all/themes/zap/pdf/Promotion-of-GB.pdf (Last accessed: 04.01.2023)

Raworth, K. (2017). *Doughnut Economics: Seven Ways to Think Like a 21st-Century Economist.* Chelsea Green Publishing, Vermont.

Sastry, M. & Michael, K. (2019). *Drivers of Climate Action in Indian Cities.* Retrieved from: www.teriin.org/sites/default/files/files/Drivers-Climate-Action-Report.pdf (Last accessed: 18.01.2023)

Shrubsole, C., Hamilton, I.G., Zimmermann, N., Papachristos, G., Broyd, T., Mumovic, D., Zhu, Y., Lin, B. & Davies, M. (2018). Bridging the gap: The need for a systems thinking approach in understanding and addressing energy and environmental performance in buildings. *SAGE Journal*, 28(1). https://doi.org/10.1177/1420326X17753513

TERI (2019). *Faster Adoption of Electric Vehicles in India: Perspective of Consumers and Industry.* The Energy and Resources Institute, New Delhi. Retrieved from: https://www.teriin.org/sites/default/files/2019-11/Faster%20adoption%20of%20electric%20vehicles%20in%20India.pdf (Last accessed: 23.01.2024)

van Ruijven, B., Urban, F., Benders, R.M.J., Moll, H.C., van der Sluijs, J.P., de Vries, B. & van Vuuren, D.P. (2008). Modeling energy and development: An evaluation of models and concepts. *World Development*, 36(12): 2801–2821. https://doi.org/10.1016/j.worlddev.2008.01.011

Warren, L. (2021). How the denim deal is making recycling the new industry standard. *Sourcing Journal.* Retrieved from: https://sourcingjournal.com/denim/denim-business/amsterdam-denim-deal-recycling-standard-scotch-soda-fibersort-269692/ (Last accessed: 18.01.2023)

WCC (2021). *Action Area: Other City-Wide Initiatives.* Wellington City Council. Retrieved from: https://wellington.govt.nz/climate-change-sustainability-environment/climate-change/what-were-doing-about-climate-change/our-climate-action-areas/other-city-wide-initiatives (Last accessed: 18.01.2023)

Zhou, B., Rybski, D. & Kropp, J.P. (2017). The role of city size and urban form in the surface urban heat island. *Scientific Reports*, 7(1): Article 1. https://doi.org/10.1038/s41598-017-04242-2

Index

9781032251523